Performing Ecological Risk Assessments

Edward J. Calabrese
Linda A. Baldwin

LEWIS PUBLISHERS
Boca Raton   Ann Arbor   London   Tokyo

Library of Congress Cataloging-in-Publication Data

Calabrese, Edward J., 1946–
    Performing ecological risk assessments / Edward J. Calabrese.
Linda A. Baldwin.
      p.   cm.
    Includes bibliographical references (p.   ) and index.
    1. Ecological risk assessment.   I. Baldwin, Linda A. II. Title
QH541.15.R57C34   1993
333.7′14—dc20   92-44578
ISBN 0-87371-703-1

LEWIS PUBLISHERS
121 South Main Street, Chelsea, MI 48118

PRINTED IN THE UNITED STATES OF AMERICA
1 2 3 4 5 6 7 8 9 0

Printed on acid-free paper

Edward J. Calabrese:

To Brian and Davy

Linda A. Baldwin:

To Steve

Preface

The field of ecological risk assessment is indeed a fledgling one despite the long history of concern with environmental contamination. Observations of chlorinated hydrocarbons and their impact on birds of prey, scores of fish kills causally related to toxic agents, and high incidences of fish tumors associated with contaminated sediments in a number of major waterways have been documented for many years.

These types of observations have frequently evoked strong reactions, such as the banning or restricting of pesticide use and the desire to create environment-protecting legislation (e.g., those relating to endangered species). As a result, many of our societal activities in the field of ecological risk assessment have been in response to actual and perceived problems and/or crises.

Ecological risk assessment is also emerging as a discipline in response to the need to create ecologically- and toxicologically-defensible schemes to evaluate the impact of contaminants on the environment. The term "environment" by necessity must be broadly defined to include the abiotic and biotic realms and their interrelationships. The concept also embodies the web of biotic relationships as seen in complex food web analysis. Thus, ecological risk assessment is not limited to how a particular species is affected, but must address the complexity of factors affecting the health and vitality of the ecosystem.

The process of ecological risk assessment is not an ideological activity with philosophical, political, or social goals. It is a process by which the ecological risk assessor, like the human risk assessor counterpart, attempts to provide the risk manager with as comprehensive an assessment of the impact of contamination on the site(s) in question as possible. While the information will include descriptive and qualitative aspects, the risk manager will be markedly aided by the presence of quantitative information, calibrated and validated predictive models and procedures so that dose-response relationships can be estimated and potential and actual impacts assessed.

This type of assessment is needed because the results of the ecological risk assessments will help managers decide if, and to what extent, remediation activities can be undertaken at contaminated sites. In such cases, the judgments will wrestle over the difficult, and usually not an all-or-none, decision of what is "acceptable" residual contamination. Thus, the risk assessment procedure will be a critical factor in helping the risk manager decide what to do.

The issue today, then, is not whether ecological risk assessment is needed, but what an ecological risk assessment should consist of and how it should be conducted. As easy as this is to say, the fact of the matter is that an integrative synthesis of how to conduct a comprehensive ecological risk assessment has not been easy to find. There certainly has been much research on sediment contamination, aquatic toxicology and food web modeling to assess exposure to so-called sink species. In addition, a variety of journals have been created in the past decade, such as *Environmental Toxicology and Chemistry, Aquatic Toxicology, Ecotoxicology and Environmental Safety*, that have responded to the pressing need to provide a communication vehicle to rapidly share emerging information.

Despite the large quantity of research in this area, the critical role of the above-mentioned journals, the leadership response and recommendations of the U.S. Environmental Protection Agency (EPA, 1992, 1992a, 1992b) and experts in the European community (Okkerman et al., 1991) no source seems to integrate the theory and practice across the range of concerns (e.g., sediment/soil, aquatic/terrestrial life, exposure assessment, etc.). It was in response to this need that the present work was initiated.

This book, therefore, attempts to compile the work of others so that the various aspects comprising an ecological risk assessment are under one roof, so to speak. This referenced compilation is designed to provide not only critical direction, but also to save time locating critical information. In addition to physically linking the work of leading researchers and assessors under one roof, this work also attempts to provide the authors' perspective of what is an ecological risk assessment, how it compares with its human counterpart, and how it should be conducted. The book recommends how issues of uncertainty can be addressed, what endpoints should be considered, how the various media and biota can be linked and applied to site-specific assessments, and how endangered species should be considered as compared to nonendangered species, among other concerns.

We hope that this work will be seen as helpful to those attempting to conduct an ecological risk assessment. The book is designed to aid those performing such tasks not only to operationally carry out the task, but also to understand why they are following the technical pathways selected.

In addition, we expect the book to be intellectually challenging on many specific fronts, as we have attempted to establish a framework for what we believe is the most toxicologically defensible way to proceed. We do not expect all to agree; in fact, we expect that our own views will continue to be modified with both further reflection and the emergence of more data. Thus, we recognize that this will not be the final word on this topic, nor should it be. We hope it will be of practical use to those who must conduct assessments and helpful to those formulators of ecological risk assessment theories as the field continues to evolve.

HOW THIS BOOK IS ORGANIZED

There will be many details presented in the pages that follow, and it may be easy to get a little lost and confused. However, at the start of this journey concerning how to conduct an ecological risk assessment it is necessary to have a clear sense of where this book is headed and why. We will lead the reader through three consecutive and interconnected boxes, if you will. Each box represents a separate major component of the ecological risk assessment process. The boxes are:

BOX 1	BOX 2	BOX 3
Deriving a Chemical-Specific, Species-Specific MATC	Deriving a Chemical-Specific Ecosystem MATC	Deriving a Multi-Contaminant Ecosystem MATC

The goal, of course, is to proceed in a proper manner through each box, thereby completing the ecological risk assessment. We believe that the ecological risk assessor must be quite clear on what his or her specific goals are for each process (i.e., box), and must follow the proper sequences for completing each segment. We believe that there will be great opportunity to challenge and debate many of the specific recommended courses of action within each box. However, we feel that the reader needs to have a clear blueprint of where he or she is heading.

Chapters 2 through 7 are devoted to providing the means and justification for deriving a chemical-specific, species-specific MATC. This task clearly dominates about three-fourths of the book and is embodied in Box 1.

Chapter 2 presents an introduction to environmental fate models and to pharmacokinetic factors as they affect exposure parameter estimations in ecological risk assessment. The critical role that pharmacokinetic and modeling factors play in ecotoxicology and ecological risk assessment is evident in Chapter 3 when the concept of the maximum acceptable tissue concentration (i.e., MATissueC) is introduced along with its current applications at hazardous waste sites. The MATissueC ecological risk assessment methodology is widely touted for use with contaminants having high fat solubility and long persistence.

Chapter 4 builds directly on the foundation laid in Chapter 3 on uncertainty factors (UFs) by providing extensive documentation and critique for the quantitative basis of UFs in ecological risk assessment. This information offers ecological risk assessors an improved foundation upon which to derive UFs in their respective assessments.

Chapter 5 presents a systematic scheme for deriving chemical-specific, species-specific MATCs based principally on the work of Barnthouse et al. (1990), but includes the use of the UF framework developed in the previous chapters. This chapter is founded mainly on the principles and applications of aquatic toxicology and has a clear focus on the aquatic environment. However, the principles, in general, are equally applicable to aquatic and terrestrial species. An appendix of several examples was also developed to show how the methodology can be specifically applied (see Appendix 2).

Chapter 6 shows how the information on UFs from Chapter 5, and the modeling approaches of Barnthouse et al. (1990) in Chapter 6 can be applied to the derivation of the equivalent of an MATC for terrestrial animals. We have adopted the term terrestrial reference value (TRV) for this purpose. This chapter also provides a critique of the MATissueC approach versus the TRV methodology for application in ecological risk assessments.

Chapter 7 directs its focus on the issue of sediment quality criteria (SQC). The chapter emphasizes a description of the principal approaches for SQC derivation, how they are conducted, and what their strengths and limitations are. How this information relates to the concept of species-specific, chemical-specific MATCs, as well as chemical-specific ecosystem MATCs, is provided.

Chapters 2 through 7 are principally directed toward one main goal: helping to derive a chemical-specific, species-specific MATC. Only after this complex process is completed can the next task of how to derive chemical-specific ecosystem MATCs (Box 2) be undertaken. Chapter 8 identifies and critiques various approaches, including those offered by the U.S. EPA and the Netherlands, that attempt to derive chemical-specific ecosystem MATCs. All previous chapters dealt with single chemical agents for which a MATC was to be derived. The co-presence of multiple agents was not addressed. However, Chapter 9 provides the means to derive multi-contaminant ecosystem MATCs. This is an area where little specific guidance has yet to be offered by various national groups. Finally, Chapter 10 offers a perspective on the strengths and limitations of current approaches in ecological risk assessment and recommendations on how the field needs to evolve to provide a more secure foundation for improved ecological risk assessments.

Edward J. Calabrese is a board certified toxicologist who is professor of toxicology at the University of Massachusetts School of Public Health, Amherst. Dr. Calabrese has researched extensively in the area of host factors affecting susceptibility to pollutants, and is the author of more than 270 papers in scholarly journals, as well as 23 books, including *Principles of Animal Extrapolation; Nutrition and Environmental Health*, Vols. I and II; *Ecogenetics; Safe Drinking Water Act: Amendments, Regulations and Standards; Petroleum Contaminated Soils*, Vols. 1, 2, and 3; *Ozone Risk Communication and Management; Hydrocarbon Contaminated Soils*, Vols. 1 and 2; *Hydrocarbon Contaminated Soils and Groundwater*, Vols. 1, 2, and 3; *Multiple Chemical Interactions; Air Toxics and Risk Assessment; Alcohol Interactions with Drugs and Chemicals, Regulating Drinking Water Quality, Biological Effects of Low Level Exposures to Chemicals and Radiation, Contaminated Soils: Diesel Fuel Contamination, Risk Assessment and Environmental Fate Methodologies,* and *Principles and Practices for Petroleum Contaminated Soils.* He has been a member of the U.S. National Academy of Sciences and NATO Countries Safe Drinking Water committees, and of the Board of Scientific Counselors for the Agency for Toxic Substances and Disease Registry (ATSDR). Dr. Calabrese also serves as Chairman of the International Society of Regulatory Toxicology and Pharmacology's Council for Health and Environmental Safety of Soils (CHESS) and Director of the Northeast Regional Environmental Public Health Center at the University of Massachusetts.

Linda A. Baldwin is currently a research assistant in the Environmental Sciences Program at the University of Massachusetts in Amherst. She received her BS in Biology from the State University of New York College at Oswego. Ms. Baldwin has been involved in both human and ecological risk assessment. Her research interests and publications are in the areas of aquatic toxicology and health effects associated with pollutants.

Contents

List of Figures

List of Tables

Performing Ecological Risk Assessments

1

Ecological Risk Assessment:
An Overview

This book represents a guidance document addressing the factors necessary for consideration in conducting an ecological risk assessment, their toxicological foundations, and the operational procedures used in the estimation of acceptable exposure levels. Included in the spectrum of ecological exposure criteria are methodologies for how to derive acceptable exposures for aquatic and terrestrial species, and acceptable toxicant concentrations in water, soils, and sediments. In addition, this report provides rationales and procedures for deriving more broadly based ecosystem exposure criteria.

While each of these environmental receptors (e.g., aquatic and terrestrial life, water, soils, sediments, and ecosystems) are intimately related, the derivation of their respective environmental criteria has generally had its own evolutionary history stimulated by different legislative initiatives and under the leadership of diverse segments of the research and regulatory communities. Consequently, even though ecological risk assessment is a relatively new field, its components have been evolving at considerably different rates, with highly variable supportive databases in terms of quantity and quality. For example, the development of criteria to assure the protection of aquatic life has an especially long history with experimental protocol. These criteria, while constantly evolving, are both highly refined and well understood. On the other hand, development of risk assessment procedures for terrestrial organisms is a new concern with more limited experience to draw upon.

As a consequence of the differential development of the diverse, but inter-related, aspects of the emerging field of ecological risk assessment, a set of descriptive terms has been formulated that is usually quite helpful, but can be confusing and redundant. For example, in the aquatic domain, there has been a long history of deriving what is called the maximum acceptable toxicant concentration (MATC). However, in recent years the EPA recommended that the term be changed to the final chronic value (fCv) (Stephan et al., 1985). Although this term has been subsequently cited (Norberg-King, 1989; Suter, 1990; DiTorro et al., 1991), the term MATC is still widely used. More recently, Fordham and Reagan (1991) have proposed the use of a new term entitled the maximum acceptable tissue concentration, which they call the MATC. If this looks confusing, it is, since the original MATC deals with environmental concentrations, not tissue levels. Consequently, we propose that the Fordham and Reagan (1991) MATC be slightly modified to MATissueC to avoid confusion. The MATissueC can be used for both aquatic and terrestrial organisms and will be in this report. Contractors for the Department of Defense (HLA, 1991) working at the Rocky Mountain Arsenal (RMA) have also derived the term terrestrial reference value (TRV) to describe an acceptable contaminant exposure rate for terrestrial animals. This TRV is purported to be very similar to the term reference dose (RfD), which is a standard term used in human risk assessment. Consequently, the reader must be somewhat nimble in adapting to a flux of new terms describing both old and new concepts.

Of particular relevance is the need to appreciate the massive effort expended in human risk assessment methodology development, and how it can be useful to the process of ecological risk assessment. Equally important is understanding where the goals of the two approaches differ. In assembling material for this document, the lack of adequate communication between the fields of ecological and human risk assessment was striking. This book has been written with the intention of keeping the lines of communication open between the two areas, while at the same time being careful to delineate the legitimate overlapping and distinct goals of each endeavor. The principal point, however, is that professionals of both domains have a lot that they can learn from each other. We hope this report fosters that in word and in deed.

As noted in the opening paragraph, this book is designed both to address the toxicological foundations for ecological risk assessment and to perform such exercises. Thus, the work provides the rationale for why certain approaches are used, the experimental data supporting those approaches with adequate reference citation, and the computational procedures to derive quantitative estimates. Thus, it is intended that the reader will be able to link theory and its documentation with appropriate follow-through to practical examples.

As in the case of human risk assessment, the process of ecological risk assessment is comprised of four components: hazard assessment, exposure assessment, dose-response estimation, and risk characterization. While each of these components is conceptually the same in the two types of risk assessment and actually drive each process, they are used to help achieve the different goals of the respective approaches. For example, the human risk assessor tries to provide risk managers with a comprehensive assessment of the response of the population at each level of exposure over the normal life span. Nestled within this perspective is a concern for the identification of factors affecting susceptibility, including quantifying high risk groups, both in terms of number of individuals affected and degree of toxicity enhancement. Thus, there is the strong intention of describing not only how the population may respond in general, but how high risk segments may be affected. The human risk assessor is also interested in a wide range of toxicological endpoints from subtle biochemical changes to frankly toxic responses.

The ecological risk assessor, while generally following the same procedural pathway, focuses on defining the impact of exposures on the population in terms of growth, maintenance, and reproduction. As a result, ecological risk assessment usually only considers high risk groups as part of a sensitive life-stage evaluation. Therefore, the general distinction is that human risk assessment has a focus on the individual, while ecological risk assessment has its focus on the population.

While this individual versus population distinction sounds, and in practice is, generally clear, the situation becomes much less distinct when the issue of protecting endangered species and migratory birds is concerned. In these cases, Congress has determined that individuals, not just populations, will be protected. Ecological risk assessment, in these instances, will resemble its human counterpart.

Each risk assessment concept and extrapolation procedure of relevance to the field of ecological risk assessment is presented and critiqued in this report. The methodologies that purport to offer guidance for conducting ecological risk assessments are then assessed in reference to these critical concepts. Based on these evaluations, recommendations are made using toxicologically defensible principles for the derivation and selection of an appropriate extrapolation procedure. Finally, the document identifies the range of supportive databases for each extrapolation procedure, and presents toxicologically based methodologies for deriving species-specific and ecosystem exposure criteria.

This document has attempted to assess the strengths and limitations of each procedure, their respective databases, and the general confidence and reliability of the various options available at each step in the process of ecological risk assessment. Consideration is also provided to the role of professional judgment and limited weight of evidence schemes in deciding on particular courses of action. It is believed that the proposed methodology provides the risk assessor with a sound foundation upon which reliable judgments can be made. This includes a consideration of the limitations of supportive databases and the flexibility to select among competing options based on quality of data, biological endpoints, statistical methods, resource limitations, and environmental protection philosophy.

This book, therefore, has relevance to broadly based ecological risk assessments, including those of an aquatic and terrestrial nature. The procedures recommended can be applied to a wide range of ecological settings, regardless of their complexity.

Part I

Concepts and
Theoretical Foundations

2

Exposure Assessment

A. ENVIRONMENTAL FATE MODELING

The process of risk assessment, regardless of whether it involves human or ecological receptors, includes an exposure assessment component. With respect to human risk assessment, the Environmental Protection Agency (EPA) has sponsored the development of an exposure assessment handbook that details the relevant factors associated with contaminant exposures. However, the issue of exposure assessment within the context of ecological risk assessment is more complicated because of the varied nature of the receptors (i.e., multiple species at different trophic levels) and the generally site-specific nature of the assessment.

The process for hazard evaluation of environmental toxicants has been assessed by Maki and Bishop (1985). The first step in this process to provide quantitative estimates of exposure involves the identification of the physical and chemical properties of the agents of concern, which includes an estimation of the amount of the contaminants in the defined environment. Based on the physical and chemical properties of the compounds, as well as the estimated quantities and areas of chemical loadings, the critical step is to predict the concentrations of the chemical agents in various environmental compartments. These estimates are useful in several respects in that they may assist in establishing priorities for chemical agents of concern, as well as for projecting species exposures.

Environmental fate modeling is often employed to organize chemical and environmental data into an objective mathematical description of the behavior of chemical agents in aquatic and terrestrial ecosystems (Burns and Baughman, 1985; Clark et al., 1988). When the information is supplemented with data of potential or actual

7

releases of chemicals to water bodies, the estimated exposure concentrations, magnitudes of the fate processes, and persistence of the chemical can be estimated in a quantitative manner. The linkage of this information with the results of toxicology studies on the receptors of concern is of direct assistance in assessing the risk posed by chemical releases to the environment. Thus, it is hoped that fate models can be employed to estimate the site-specific concentrations of the parent and transformed chemical contaminants. These concentrations can then be compared to chemical concentrations that toxicological validations have determined cause harm to aquatic/terrestrial animals.

The use of environmental fate modeling results in the development of behavioral profiles of chemicals in a specific environmental setting. The predictions of chemical behavior may be directed to the persistence of the agent, as well as dominant fate processes, the magnitude of chemical exposures, and the relative distribution of the residual chemical concentration. Of particular relevance is the work of Mackay and colleagues (1985) which derived a series of general evaluative models to characterize the behavior of synthetic chemicals in the biosphere. The principal concept centered on the capacity of chemicals to accumulate in different environmental sectors (air, soil, water, animals, etc.). This escaping from one medium to another is deemed "fugacity" and can be related to environmental concentrations for model predictions.

The fugacity calculation has broad application with relevance to aquatic systems as exemplified in the exposure analysis modeling system (EXAMS), which was designed principally to assist in the identification and characterization of agents as potential aquatic pollutants. Such a model requires inputs on the chemistry of the agents of concern (e.g., hydrolytic rate constant), characterization of the aquatic and terrestrial environment (e.g., pH, temperature, hydrologic input, etc.), and chemical loadings into the system.

Based on such information, the model provides estimates of expected environmental concentrations, the fate of the agent(s), and the persistence of the agent(s) in that system. However, as noted by Burns and Baughman (1985), the modeler must be acutely aware of the potential of agents to undergo changes, such as ionic speciation within the modeled system and complexation reactions. For example, the LC_{50} for hydrogen cyanide (HCN) in juvenile fathead minnow decreases by 90% as the pH of the water increases from 6.5 to 9.3.

Use of environmental fate modeling is crucial to properly estimate the site-specific behavior of agents of concern. It is important to note that current contaminant transport models have a unique range of limitations. In general, they often lack adequate data on sediment-water interactions, as well as having inadequate model verification. Verification of model predictions assessed over a different time period and under varying major environmental parameters is needed to confirm that the model incorporates major transport mechanisms both qualitatively and quantitatively.

Within the modeling process, the data requirements for fate analysis involve parameters such as water solubility, vapor pressure, octanol-water partition coefficient, bioconcentration factors (BCF), soil sorption constant, water-air ratio, and degradation rate constants in water, air, soil, and biota.

Water solubility and vapor pressure are critical in fate analysis since they may be employed to estimate other critical factors such as the BCF, octanol-water partition coefficient, and soil sorption constant. A combining of the information derived from water solubilities and vapor pressure also assists in estimating the volatility potential of chemicals from soil and water (Mackay et al., 1985; Mackay, 1991).

The octanol-water partition coefficient and BCF are of critical environmental relevance since they may be used to predict the movement of the agents into biota. Measurement of the octanol-water partition coefficient offers an estimate of the BCF, while both correlate with soil sorption to provide a quantitative estimate for the capacity of a chemical to partition between water and soil solids. This parameter (i.e., soil sorption parameter) is required to assess leaching through soil, volatilization from soil, and volatilization from aquatic systems. When combined with an estimate of stability in soil or water, the modeler may predict whether leaching and volatilization will be significant migration pathways for agents in aquatic and terrestrial environments. The utility of the soil sorption constant for fate analysis is based on its capacity to express adsorption independent of soil type, since adsorption of many organic chemicals is related directly to the content of soil organic carbon. In the case of metals, possibly polymers and positively charged molecules, their adsorption to soil is not necessarily correlated with organic carbon content. In these cases, some other soil property may be a better predictor than organic carbon. The water-air ratio (i.e., the inverse of Henry's law constant for convenience) provides a means to estimate the partition between water and air. When used in conjunction with the soil sorption constant and an estimate of the chemical's decomposition rate in each system, it is possible to evaluate the overall potential of volatilization as an important mechanism of migration from water and soil to air (Mackay et al., 1985; Mackay, 1991).

A recent attempt to estimate ecological risks in aquatic systems based on an integrated fate and effects model (IFEM) was put forth by Bartell et al. (1988). This IFEM represents a synthesis of the Fate of Aromatics (FOAM) and the Standard Water Column models. Since the FOAM model was developed for polycyclic aromatic hydrocarbons (PAHs), naphthalene was utilized as a model contaminant (Breck and Bartell, 1988).

B. TOXICOKINETICS IN EXPOSURE ASSESSMENT

1. Bioconcentration Factor (BCF)

One of the principal concepts in ecological risk assessment is that of bioconcentration. Bioconcentration activities of organic and inorganic agents differ in that uptake of organic substances by aquatic organisms is directly related to aquatic concentrations, while the uptake of inorganic substances is influenced by homeostatic mechanisms or competitive uptake of other ions.

Most bioconcentration measurements of lipophilic compounds correlate directly with the octanol-water partition coefficient and inversely with the agent's water solubility. However, correlations between bioconcentration and physical properties are poor for very large molecules of high molecular weight and those agents that are readily metabolized by fish. It is now recognized that large molecules are not efficiently transported across the gill membranes even though they can be assimilated by fish from food (Oliver and Niimi, 1985). With respect to metabolic factors, Southworth et al. (1980, 1981) reported that BCFs for PAHs metabolized by fish are much lower than predicted from their octanol-water partition coefficient.

It should be emphasized that the shorter the chemical's half-life in the fish, the more rapidly the equilibrium concentration in the fish is established. In their study, Oliver and Niimi (1985) estimated that chemicals with half-lives of 22 days or less will achieve 95% of their steady-state concentrations by about 100 days. For compounds with long half-lives in fish, the use of laboratory BCFs (estimated either kinetically or at steady state) markedly underestimated residual levels.

2. Octanol-Water Partition Coefficient

The toxicological relevance of the octanol-water partition coefficient is strikingly seen for agents inducing narcosis as their mode of toxic action. In this case, a high correlation exists between the fish LC_{50} values and the log of the octanol/water partition coefficient (Veith et al. 1983; Smith and Craig 1983; Konemann 1980; McCarty et al. 1985). In addition to using the octanol-water partition coefficient to predict acute toxicity, McCarty et al. (1985) raised the critical question of whether it could also be used to estimate chronic toxicity using chlorobenzenes and substituted phenols in the rainbow trout. They also were interested in whether the slope of the relationship of the log octanol-water partition coefficient with chronic toxicity would be parallel with the slope seen with acute toxicity. If that were the case, it would serve to further validate the Application Factor (AF) concept (Chapter 4) and establish a role for using the chemical structure [i.e., via the process of quantitative structure activity relationships (QSAR)] to estimate chronic toxicity values.

While the original report of McCarty et al. (1985) did not reveal parallel slopes for acute and chronic response, a subsequent report by McCarty (1986) using a larger database and a more appropriate statistical approach was able to show that acute and chronic QSARs for groups of chemicals with similar modes of action are parallel with the log of the octanol-water partition coefficient as the predictor. The author argued that the relationship between toxicity and bioconcentration in their study showed that the molecular characteristics controlling toxicokinetics and factors affecting toxic action are closely related.

This conclusion must be currently restricted to chemicals with similar modes of action at both the acute and chronic levels. Generalizations are, therefore, highly limited at present. Nonetheless, such attempts to validate the concept of the AF and expand its relationship to QSAR is an important development.

C. TOXICOKINETIC APPROACHES FOR EXPOSURE ASSESSMENTS

The following section, based on the procedures offered by Spacie and Hamelink (1985), is designed to provide step-by-step approaches for exposure estimation in aquatic and terrestrial organisms based on a knowledge of the toxicokinetics of the affected organisms. The procedures will lead the reader through the use of single- and two-compartment models, including a consideration of whether growth is assumed or excluded. The models involve a direct consideration of bioconcentration in an aqueous environment, bioaccumulation, biomagnification, and food web modeling as affecting both aqueous and terrestrial organisms. The methodology will include schemes whereby data are obtained from experimentation or, in the absence of data, where and how the input parameters may be reasonably estimated.

1. Single Compartment Model Based on Experimental Data

a. Assumptions:

1. Each organism behaves as a single compartment.
2. Uptake of the contaminant from water is the sole input factor and is directly proportional to its concentration in the water (Cw).
3. All residual molecules of the contaminant in the fish are in a common pool.
4. All molecules of the contaminant in the fish are equally available for elimination (i.e., depuration).
5. Regardless of the mechanism of elimination, the rate of elimination is assumed to be first order; first order means that elimination is directly proportional to the concentration in the fish (C).
6. The size or volume of the compartment does not change over the duration of the study; this means that no growth can occur with this method.

b. Procedure: Calculation of Steady-State BCF

1. The rate of change of the contaminant in the fish is described by:

$$\frac{dC}{dt} = \text{uptake} - \text{loss} = K1\ Cw - K2\ C \qquad (2.1)$$

where

Cw = the concentration of agent in water (μg/mL)
C = the concentration of chemical in the fish (μg/g)
t = time (h)
K1 = the uptake rate constant (mL/g.h or assuming a tissue density of 1 g/mL, h-1)
K2 = the first order rate constant for depuration (h-1)

2. At the steady-state condition, the rate of uptake equals the rate of elimination, yielding.

$$\frac{dC}{dt} = O = K1\ Cw - K2Css \qquad (2.2)$$

where Css is the concentration of contaminant in the fish at steady state.

3. Based on Equation 2.2, one can solve directly for Css:

$$0 = K1Cw - K2Css$$

$$K2Css = K1Cw$$

$$Css = \frac{K1Cw}{K2} \qquad (2.3)$$

4. The steady-state BCF, directly proportional to Cw (assumption 2), may then be described as:

$$BCF = \frac{Css}{Cw} = \frac{K1}{K2} \qquad (2.4)$$

c. *Estimation of the Amount of Contaminant Taken Up Per Unit Time*

1. Integration of Equation 2.1 yields the highly useful equation below that incorporates explicitly the time component:

$$C = \frac{K1}{K2}\ Cw\ [1 - \exp(-K2t)] \qquad (2.5)$$

where exp ($-$K2t) is the base of natural logarithm (e) elevated to the power $-$K2t.

An equivalent form of Equation 2.5 is

$$C = Css \, [1 - exp \, (K2t)] \qquad (2.6)$$

This equation describes a curvilinear increase in the concentration of the contaminant in the fish followed by a plateauing such that an asymptotic characterization is achieved.

2. Estimation of the elimination rate constant (K2) can be made as follows. If an apparent steady-state is reached during the course of a continuous study, K2 can be estimated by determining the time (t50) needed to achieve 50% Css:
    50% of steady-state is achieved when C = 0.5 Css. Substituting this value into Equation 2.5 allows one to calculate the time to reach this 0.5 Css value. This is also known as the elimination half-life.

$$t50 \, = \, \frac{\ln 0.5}{K2} \, = \, \frac{0.693}{K2}$$

Solving for K2:

$$K2 \, = \, \frac{\ln 0.5}{t50} \, = \, \frac{0.693}{t50}$$

K2 is usually determined experimentally by placing exposed fish into uncontaminated water. There is no requirement that the fish need to be at steady-state when the elimination process is started. The data collected are in the form of a plot made of the reduction in C (on a log scale) against the elimination time. K2 can then be calculated from the slope of the straight line:

$$\ln C \, = \, \ln Co - K2t \qquad (2.7)$$

where Co is the contaminant level in the fish at the beginning of the elimination phase.
    Advantages of the direct estimation method for estimating K2 are (1) the data can be fit by a simple least squares regression with confidence limits, and (2) deviation from linearity will allow one to seek out more appropriate models.

3. K1 can be calculated once K2 is known, using a formula derived from Equation 2.3 with C substituted for Css:

$$C \, = \, \frac{K1 \, Cw}{K2}$$

$$\frac{C(K2)}{Cw} = K1$$

Taking into account factors relating to residual levels of contaminant in the body and rate of elimination the calculation becomes:

$$K1 = (dC/dt + K2C) / Cw \qquad (2.8)$$

Thus, in order to estimate K1 it is necessary to know K2, the concentration in the fish over time, and the concentration of the contaminant in the water over time.

2. Single Compartment Model Based on Estimated Data

The previous method was able to derive K1 and K2 values from experimental data. Unfortunately, resources and time may be too limited to obtain these values experimentally. In the absence of experimental data, Neely (1979) has provided a straightforward way to estimate K1 and K2 rate constants based on the physiology of the fish and the properties of the agent.

a. Procedure: Estimation of K1 and K2 in Absence of Data

1. A physiologically based uptake model is required, such as the following (Neely, 1979):

$$\frac{dC}{dt} = \frac{ECw\ Rv}{F} \qquad (2.9)$$

where
  E   = efficiency of agent to pass across the gill membrane
  C, Cw = concentration of chemical in fish and water, respectively
  Rv   = volume of water flowing past the gills per unit time
  F   = weight of fish

Simply put, the rate of uptake of the agent is equal to the efficiency by which the fish extracts the agent from the water flowing past the gill surface.

2. Rv is a function of the oxygen concentration in the water and the amount of oxygen needed to sustain the respiration rate of

the animal and is calculated by the following formula (Norstrom et al. 1976):

$$Rv = \frac{Q}{Eox\ Cox} \qquad (2.10)$$

where
  Q   = total energy metabolism of the fish (mL of O_2/h)
  Eox = efficiency of oxygen transfer across the gills
  Cox = concentration of oxygen in the water (mL of O_2/mL of water)

(Note that the total energy metabolism per gram of fish becomes smaller as fish grow.)

The relationship of Q and weight is provided by:

$$Q = aF^r \qquad (2.11)$$

where
  r = growth exponent of fish and estimated by Norstrom et al. (1976) to be 0.8
  a = metabolic rate coefficient in units of mL of O_2/h g$^{-0.8}$ of fish as derived by Fry (1957) and converted into units employed by Norstrom et al. (1976).
  F = weight of fish

Neely (1979) cited Lloyd (1961) as estimating Eox as 0.75. However, this value is likely to be affected by factors such as temperature and size of fish. The above procedure, therefore, indicates that if the metabolic rate coefficient (a) is known, along with the oxygen concentration in the water and the weight of the fish, then the ventilation volumetric rate Rv can be estimated.

3. E is estimated as follows. The efficiency by which an agent moves across the gill membrane is believed to be the result of a partitioning phenomenon and can be estimated from the log Kow. Based on a plotting of the average value of Eox between active and standard metabolism against the log of the Kow for selected agents (Table 2 of Neely, 1979), the following equation for Eox was derived:

$$E = [0.07 \log Kow - 0.02]$$

4. K1 is calculated from an equation formed by combining Equation 2.9 with the part of Equation 2.1 representing uptake:

$$\frac{dC}{dt} = \frac{E \, CwRv}{F}$$

$$\frac{dC}{dt} = K1Cw$$

$$K1Cw = \frac{E \, CwRv}{F}$$

$$K1 = \frac{E \, Rv}{F} \qquad (2.12)$$

Substituting for E:

$$K1 = \frac{(0.07 \, \log Kow - 0.02) \times Rv}{F}$$

5. K2 can be estimated from Equation 2.4:

$$K2 = K1/BCF$$

Kl has just been estimated. BCF can be determined by direct measurement or estimated from any of the following three equations depending on which conditions are employed (e.g., whole fish, muscle, etc.):

$$\log BCF = 0.76 \, \log Kow - 0.23 \text{ based on BCF} \qquad (2.13)$$

measured in whole fish from Veith et al. (1979).

$$\log BCF = 0.542 \, \log Kow + 0.124 \qquad (2.14)$$

based on fish muscle from Neely et al. (1974).

$$\log BCF = -0.508 \, \log S + 3.41 \qquad (2.15)$$

where S = water solubility in μmol/L
based on fish muscle from Chiou et al. (1977).

The following two examples demonstrate estimation of the kinetic constants (K1, K2) and compare the results to experimentally derived values.

Example 1: <u>Chlorpyrifos Uptake Experiment in
Aquarium Goldfish (Neely, 1979)</u>

Data:

Goldfish average weight (F)	1.2g
Temperature of water	25°C
O_2 concentration (Cox)	7.6 ppm (5.8 × 10⁻ mL of O_2/mL water)
Metabolic coefficient (a)	0.37 (value for ''active metabolism'' from Mayer and Kramer, 1973; Table 1 of Neely, 1979)
Log Kow	4.6
Growth exponent of fish (r)	0.8 (estimated by Norstrom et al., 1976)
Efficiency of oxygen transfer across the gills (Eox)	0.75

A. Estimation of K1

1. Calculate Q:
$$Q = aF^r$$
$$Q = 0.37 \ (1.2)^{0.8}$$
$$Q = 0.32$$

2. Calculate Rv:

$$Rv = \frac{Q}{Eox \ Cox} = \frac{0.32}{(0.75)(5.8 \times 10^{-3}/mL \text{ of } O_2 \text{ mL water) h} - 1}$$

$$Rv = 111 \text{ mL of water h} - 1$$

3. Calculate K1:

$$K1 = \frac{E \times Rv}{F}$$

$$K1 = \frac{(0.07 \log Kow - 0.02) \times 111 \text{ mL of water h} - 1}{1.2 \text{ g}}$$

$$\frac{[(0.07)\ (4.6) - 0.02] \times 111 \text{ mL of water h} - 1}{1.2 \text{ g}}$$

$$K1 = 28 \text{ mL g} - 1 \text{ h-1}$$

This value contrasts with 55 mL g −1 h-1 as determined experimentally (Blau and Neely, 1975).

B. Estimation of K2

$$BCF = \frac{K1}{K2}$$

$$K2 = \frac{K1}{BCF}$$

K1 = 28 ml g − 1 h − 1 as determined above.

BCF is derived from Equations 2.13 and 2.15, to yield a K2 value of 1.5×10^{-2} h − 1 which was lower than the 7.8×10^{-2} h − 1 value experimentally derived (Blau and Neely, 1975).
    The basis of these discrepancies is that the experimentally derived data include the clearance of chlorpyrifos and the rate of metabolism and clearance of its metabolite which is more rapidly eliminated than the parent compound.

Example 2: TCDD Uptake Experiment in Rainbow Trout
(Neely, 1979; Neely and Blau, 1977)

Data: Rainbow Trout weight (F)   35g
      Temperature of water   13°C
      O_2 Concentration (Cox)   10.5 ppm (7.98×10^{-3} mL
                                    of O_2/mL of water)
      Water solubility of TCDD   200 ppt
      Kow for TCDD   6.19
      Growth exponent of fish (r)   0.8 (estimated by Norstrom
                                    et al., 1976)

Metabolic coefficient (a)   0.18 (interpolated from
                                 Mayer and Kramer 1973;
                                 Table 1 of Neely, 1979)

Efficiency of oxygen across
   the gills (Eox)   0.75

A. Estimation of K1

   1. Calculate Q:

$$Q = a\ F^r$$
$$Q = (.18)\ (35)^{0.8}$$
$$Q = 3.09$$

   2. Calculate Rv:

$$Rv = \frac{Q}{Eox\ Cox}$$

$$Rv = \frac{3.09}{(0.75)\ (7.98 \times 10^{-3}\ \text{mL of O}_2/\text{mL of water})\ h - 1}$$

Rv = 535 mL of water h-1 (ventilation volumetric rate)

   3. Calculate K1:

$$K1 = \frac{E \times Rv}{F}$$

$$K1 = \frac{(0.07\ \log\ kow - 0.02) \times (535\ \text{mL of water}\ h - 1)}{35g}$$

K1 = 6.3 mL g-1 h-1

This value is higher than the experimental value of 4.58 mL
g − 1 h − 1 (Neely, 1979).

B. Estimation of K2

$$BCF = \frac{K1}{K2}$$

$$
\begin{aligned}
K2 &= (K1)/BCF \\
\text{Log } BCF &= [(0.76) \ (6.18){-}0.23] \\
\text{Log } BCF &= 4.45 \\
BCF &= 28{,}183.8 \\
K2 &= \frac{6.3 \ \text{mL/g} - 1 \ \text{h} - 1}{28{,}183.8} = 2.23 \times 10^{-4} \ \text{h} - 1
\end{aligned}
$$

This value should be compared with the experimental value of 5×10^{-4} h $- 1$.

A principal assumption of the single compartment model was that no growth in the organism occurred. For agents with large partition coefficients such as DDT, dieldrin, and PCB, this is often an untenable assumption. This is because these agents approach the steady-state condition quite slowly.

As a result of the slow time to steady-state, the results of long-term bioconcentration tests with such agents often need to be adjusted in order to account for growth. The act of growth will result in a reduction of the concentration of the residues.

Spacie and Hamelink (1985) incorporated growth as a loss term in the basic kinetic model previously employed.

$$\frac{dC}{dt} = K1Cw - K2C - gC$$

Growth is assumed to be an exponential function.

$$\frac{dW}{dt} = Wo \ exp \ (gt)$$

where Wo is the weight at the start of the growth period or experiment.

g (d -1) is the growth rate constant

The two loss terms K2 and g are often combined such that K3 = K2 + g.

This will yield the following integrated model

$$C = \frac{K1 \ Cw \ [1\text{-}\exp(-K3t)]}{K3}$$

3. Two Compartment Model

Many tissue residues do not behave as being from a common pool, with some proportion of the residue being eliminated much more quickly than the remainder. This has led to the concept of fast and slow compartments and the use of a two-compartment model. In this case, uptake and elimination occur only in the fast or central compartment, while residues in the slower compartment increase as a function of the concentration in compartment 1 (i.e., fast compartment) (Spacie and Hamelink, 1985).

The classic elimination curve illustrating fast and slow phases is given as shown in Figure 2.1.

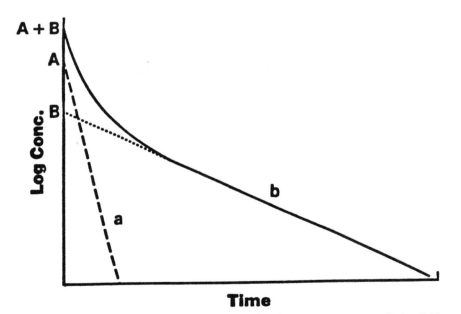

Figure 2.1 The two-compartment model. *Source:* Spacie and Hamelink, 1985.

The equation describing the two-compartment model is:

$$C = A \exp(-at) + B \exp(-bt)$$

where A and B are intercepts of the lines with slopes a and b, respectively; they are plotted logarithmically.

the quantity A + B is the amount of agent in the body at the start of elimination (Co).

The two response lines are solved by:

(a) employing a linear regression on the later part of the slow phase
(b) after B and b are determined, the remaining line can be determined by subtracting the calculated values at C at each observation time from the experimental points during the early period.
(c) A plot of the differences will yield a straight line with slope a and intercept A.

The two-compartment model requires more data points, especially during the latter phases of elimination. This can put considerable constraint on limited resources. However, the additional expenses are often justified by the improved accuracy of the model (Spacie and Hamelink 1985).

4. Bioaccumulation Model

Estimation of tissue concentrations by the BCF in the laboratory has been underestimated by an order of magnitude or more than observed values in fish in large reservoirs or lakes. This has been explained on the basis that the uptake rate via the gills at low water concentrations is small relative to the uptake rate from ingestion of chemicals in the diet.

Recognition of this situation has led to the derivation of bioaccumulation and food chain transfer models. In the case of the bioaccumulation model, a new uptake term is incorporated into the one-compartment model. The term (aRCf) involves three components:

a = the assimilation efficiency of the agent from food (g absorbed divided by g ingested).

R = weight specific ratio for feeding rate (g ingested per g of body weight/d)

Cf = concentration of contaminant in food

The bioaccumulation factor (N) is defined as:

$$N = (K1/K2 + aR/K3) \; Cf/Cw$$

This model assumes additivity of residue accumulation from food and water, and that the elimination rates of residuals derived from food and water are first-order and in the same compartment.

The data on which to base a and R can be experimentally derived. However, more likely than not, these values will be estimated based on literature values depending on the contaminants, the species, and the range of environmental conditions considered appropriate.

5. Food Chain Transfer Model

The food chain transfer model represents an expansion of the bioaccumulation model since it predicts the transfer of a chemical through an aquatic, as well as terrestrial, food chain across various trophic levels (e.g., phytoplankton, zooplankton, small fish, large fish, etc.). The concentration of the chemical of concern for each trophic level (i) is estimated by the bioconcentration factor for that level (referred to as Niw) plus the chemical contribution via feeding on the next lower level (i-l). Thomann (1981) has established, therefore, that the bioaccumulation at any trophic level can be determined once the bioconcentration factors (Niw) at each level, and the food chain transfer values (fi) for the consumers are known. Values of fi which are greater than 1.0 result in the phenomenon of biomagnification.

In this 1981 report, Thomann estimated the biomagnification of PCB residues in a four-step food chain (Table 2.1) using estimated (not measured) rate constants for K1, K2, and a. The estimated bioaccumulation factors were within a factor of 10 of the residues observed in the field.

According to Spacie and Hamelink (1985), modeling food chain transfers can be quite sensitive to the relative value of a and K3.

Table 2.1.   Sample Calculation of Food Chain Transfer from Parameters for PCB Estimated by Thomann (1981)[a]

Parameter	Trophic Level			
	1	2	3	4
k_d, $1/d$	—	0.01	0.004	0.001
g, $1/d$	—	0.0092	0.0048	0.0018
R, $g/g \cdot d$	—	0.105	0.017	0.009
α	—	0.9	0.9	0.9
N_W	$10^{5.5}$	$10^{5.29}$	$10^{4.59}$	$10^{4.9}$
N	$10^{5.5}$	$10^{6.24}$	$10^{6.49}$	$10^{6.95}$
Magnification (N/N_W)	—	$8.9\times$	$79\times$	$112\times$

[a]Sample calculation for level 3: $N_3 = N_{3W} + f_3 N_{2W} + f_3 f_2 N_{1W}$; $N_3 = 10^{4.59} + (1.74)10^{5.29} + (1.74)(4.92)10^{5.5}$; $N_3 = 10^{6.49}$; where $f_3 = \alpha_3 R_3/(k_d + g) = (0.9)(0.017)/(0.004 + 0.0048) = 1.74$.

A relatively minor error in either of their estimates can affect the predicted outcome significantly, particularly at the higher tropic levels.

The mathematical expression of the food chain transfer model is as follows:

$$Ni = Niw + \frac{ai,i\text{-}1Ri,i\text{-}1}{K3i} \frac{Ci\text{-}1}{Cw}$$

Note that Niw is the BCF at a specific trophic level.

$$Ni = Niw + \frac{ai,i\text{-}1R,i\text{-}1Ni\text{-}1,w}{K3i}$$

where
   Ni-2,w = BCF for trophic level i-1
   $fi = \dfrac{ai,i\text{-}1Ri,i\text{-}1}{K3i}$ thus
   Ni = Niw + fiNi-1

Thomann (1981) and Spacie and Hamelink (1985) applied the above scheme to four trophic levels

$$\text{Level 1 } N1w = \frac{K1}{K3} \text{ (no feeding)}$$

Level 2 N2 = N2w + f2Niw
Level 3 N3 = N3w + f3N2w + f3f2 N1w
Level 4 N4 = N4w + f4N3w + f4f3 N2w + f4f3f2N1w

The overall significance of biomagnification in exposure assessment is believed to be limited to organic compounds with high octanol-water partition coefficients (i.e., $\geq 10^{-5}$). As a practical matter, it appears that most chemicals are not biomagnified (Macek et al., 1979). However, agents such as DDT with high assimilation and low depuration rates are exceptions and can be appreciably biomagnified in slow growing, long-lived species (Moriarty and Walker, 1987; Moriarty, 1985).

D. PREDICTING TOXICITY FROM EXPOSURE MODELING/PREDICTING ENVIRONMENTAL EXPOSURES FROM INTERNAL CONCENTRATIONS

The estimation of the maximum acceptable tissue concentration (MATissueC, see Chapter 3) in aquatic species presents similar challenges, as seen with terrestrial organisms for the derivation of the acceptable environmental media concentration. The key feature is the determination of what MATissueC is considered acceptable, and how it is determined. In order to achieve this goal it will be theoretically necessary to assess the response of the species under study to standard chronic bioassay evaluation over a broad dosage range. Findings from such studies, if conducted at steady-state conditions, provide the type of database necessary for a valid consideration of the MATissueC approach.

This procedure leads to predictive relationships in which tissue concentrations may be estimated from measurable independent variables such as environmental concentration, toxic endpoint determination, and BCF (Kb). For example, since both acute LC_{50} and Kb values vary with Kow, LC_{50}s may be used to derive Kb values (McCarty et al. 1985). Based on the relationships of either Veith et al. (1979) or Mackay (1982) for Kb and Kow (McCarty et al. 1985), and the predictive equations of Konemann (1980) or Smith and Craig (1983) for LC_{50} and Kow, McCarty et al. (1985) proposed that:

$$\log Kb = -\log LC_{50} + \log K$$

$$\log LC_{50} = \log Kb + \log K, \text{ that is}$$

$$Kb = K/LC_{50} \text{ or } LC_{50} = K/Kb$$

where
  K is a constant
  LC_{50} is the water contaminant in mol/L

This equation reveals that bioconcentration strongly dominates predictions of acute toxicity in fish *if* variations in inherent toxicity of the chemical are small when compared to variations in bioconcentration. Since this equation is consistent with predictions of chronic toxicity, McCarty et al. (1985) offered a more general equation relating toxicity and bioconcentration:

$$\log Kb = A (-\log \text{toxicity}) + (\log K)$$

or

$$\log \text{toxicity} = A' (-\log Kb) + (\log K)'$$

That is:

$$Kb = K/(\text{toxicity})^A$$

or

$$\text{toxicity} = K'/(Kb)^{A'}$$

where K, K', A and A' are constants; $A' = 1/A$; $(\log K') = (\log K/A)$; and toxicity is expressed in mol/L. When A is approximately 1, these equations simplify to those presented earlier.

These mathematical relationships indicate that both the acute and chronic toxic effects of certain waterborne agents may be directly related to the quantity of the agent in the body.

The scheme put forth by McCarty et al. (1985) is consistent with the operational assumptions inherent in the conduct of aquatic bioassays. Under these circumstances, toxicity is a function of steady-state concentration, with the exception of a relatively insignificant kinetic phase prior to reaching steady-state. In this case, toxicity is believed to be a function of the inherent toxicity of the agents, as well as the external toxicant concentration and exposure time.

While toxicity is often expressed during the steady-state phase of exposure, toxicity may occur for many agents primarily during the kinetic phase of contaminant accumulation (i.e., before steady-state is achieved). In such cases, McCarty et al. (1985) indicate that toxicity is more related to the rate of net bioconcentration rather than to the inherent toxicity of the agent.

The concepts of kinetic phase and steady-state toxicity have broad theoretical and practical ecotoxicological implications. For example, because toxicity is a function of both the amount of chemical taken up and accumulated, as well as its inherent toxicity, agents with higher log Kow values may not accumulate enough in four days (96 hr. LC_{50} test) to accurately predict the toxic responses seen in experiments of longer duration. Thus, the toxicity of agents with high log Kow may be underestimated in 96 hour tests since these agents would be taken up into the organism at a different (presumably lower) rate than agents with low Kow values. It is predicted that agents with high Kow values would produce reduced toxicity in short-term studies. Based on this information, McCarty et al. (1985) concluded that assessment of high and low log Kow agents in the 96 hour LC_{50} toxicity bioassay compares the toxicity of markedly different quantities of toxicants present in the organism and not the inherent differences in chemical toxicity.

It is important to note that this assessment of aquatic toxicity assumes that bioconcentration factors are constant and do not vary with different external water concentrations. In addition, it also assumes that the time to achieve equilibrium at any given internal concentration is constant (McCarty et al. 1985). However, this series of assumptions has been challenged by McCarty et al. (1985) with data on selected chlorobenzenes derived from Oliver and Niimi (1985). These authors reported that the time to steady-state for these chlorobenzenes increased as exposure concentrations decreased.

Based on this analysis, McCarty et al. (1985) offered the following practical insight that the steady-state term of the toxicity equation dominates the toxicity for agents with log Kow values up to 3.5. In this case, the accumulation to the steady-state level proceeds quickly as compared to the duration of the bioassay. For agents with log Kow values greater than 3.5, it is predicted that accumulation may increasingly limit the toxicity occurring during the kinetic phase and may dominate the overall toxicity. McCarty et al. (1985)

concluded that acute and chronic bioassays of agents with log Kow values >3.5 (e.g., endrin Kow = 4.56; p,p' DDT Kow = 5.75; p,p' DDE Kow = 5.69) (Connell and Schuurmann, 1988) will likely underestimate toxicity, and that water quality guidelines obtained from such investigations will not be sufficiently protective.

The solution to this ecotoxicological problem, according to McCarty et al. (1985), may be found in the standardization of the exposure by basing it on internal concentrations and by addressing both the kinetic and steady-state phases of exposure. How this "solution" would be realized requires further development. However, McCarty et al. (1985) suggested that one could measure or estimate internal concentrations based on predetermined uptake (K1) and depuration (K2) rate constants. Exposure or dosage could then be given as the sum of periodic products of internal concentration and time up to a series of predetermined summation values (i.e., the integration of the area under the curve of a time versus body concentration plot).

In a subsequent report, McCarty (1986) estimated the internal toxicant concentration at a given biological endpoint (e.g., acute and chronic toxicities) based on the type of predictive equations noted above for a limited number of chlorobenzenes and substituted phenols principally in the fathead minnow and the rainbow trout. Despite an acknowledged limited database to support this proposed scheme, McCarty (1986) cited three examples where internal doses (i.e., for alkyl esters of o- and p-aminobenzoic acid, and pentachlorophenol) had been reasonably estimated based on this proposed scheme (Mackay and Hughes, 1984; Spehar et al., 1985; Kishino and Kobayashi, 1980; Kobayashi and Kishino, 1990).

Van Hoogen and Opperhuizen (1988) have recently provided additional support to the model prediction of McCarty (1986). They established that for agents with relatively high uptake and elimination rate constants, such as chlorobenzenes, the lethal toxic concentrations in water conform to the Ferguson Principle of thermodynamic equilibrium between the exposure concentrations and internal concentrations. As a result, the aqueous exposure concentration may be an effective toxicity parameter. Table 2.2 provides the relationship of aqueous concentrations of three chlorobenzenes and their concentrations in guppies at the observed LC_{50} value in relation to duration of exposures. The data reveal that the concentrations in fish at the time of death displayed little variation among the agents. Furthermore, the internal concentration

Table 2.2.   Concentrations in Water and in Guppies in LC$_{50}$ Experiments and LC$_{50}$ Values Obtained from the Literature

Compound	Exp. Concn. (μmol/L)	Concn. in Fish (μmol/g)	Exp. Time (d)
1,2,3-Trichlorobenzene	55.9 $\pm$1.8	2.71$\pm$0.74	0.1
	3.78$\pm$0.9	2.02$\pm$0.69	1
	1.92$\pm$0.4	2.38$\pm$0.71	4
	12.8	—[a]	14
1,2,3,4-Tetrachlorobenzene	1.69$\pm$0.15	2.31$\pm$0.64	4
	1.13$\pm$0.12	2.64$\pm$0.41	8
	5.12	—[a]	4
	3.71	—[a]	14
Pentachlorobenzene	0.54$\pm$0.08	2.54$\pm$0.59	4
	0.40$\pm$0.12	2.11$\pm$0.39	8
	—[b]	—[a]	4
	0.70	—[a]	14

Source: Van Hoogen and Opperhuizen, 1988.
Concentrations in each guppy were determined immediately after the fish died in the continuous-flow system.
[a]No concentration determined.
[b]No toxic aqueous concentration found that caused 50% mortality in the population.

in dead fish was similar for the different test compounds (i.e., 2.5 μmol/g) despite using markedly different exposure concentrations. These findings are consistent with the hypothesis that the guppies died when a lethal internal concentration of the chlorobenzenes was achieved.

Based on the above stated assumptions and data, the authors utilized the model proposed by Spacie and Hamelink (1985), shown in Equation 2.1, that uptake and clearance between fish and water is a first-order exchange process:

$$\frac{dC}{dt} = K1Cw-K2C$$

where K1 and K2 are uptake and elimination rate constants and C and Cw are the concentrations of the test compound in fish and water, respectively.

The integration of Equation 2.1 yields:

$$C = Cw \frac{K1 \ (1\text{-e-K2t} + C(o)e\text{-K2t})}{K2} \qquad (2.16)$$

Since it is assumed that the organism dies when the concentration achieves a specific level (e.g., chlorobenzene cases), all combinations of exposure time and exposure concentration can be normalized via Equation 2.16. Van Hoogen and Opperhuizen (1988), therefore, argue that when the lethal concentration in fish is achieved for a given proportion of the study population (e.g., 50%) at time t, the actual aqueous exposure concentration of the test chemical is the lethal concentration for that proportion of the population (e.g., LC_{50}) for that particular exposure time. Based on Equation 2.16, these authors presented a mathematical relationship between LC_{50}, exposure time, and critical internal concentration in Equation 2.17.

$$LC_{50} \text{ at time } t = \frac{C}{(K1/K2)(1^{-e\text{-}k2t})} \qquad (2.17)$$

Equation 2.17 assumes that fish will die when the lethal concentration is achieved regardless of exposure time or exposure condition. This equation is useful since it permits the estimation of LC_{50} values for any given exposure time based on measured C, regardless of exposure conditions. The equation also permits the estimation of the time to death of 50% of the population for any exposure concentration. If C is constant, the findings of static, semistatic, and continuous exposure experiments can be compared if either the exposure conditions are known or the internal lethal concentrations are reported.

The data indicating that the three chlorobenzenes display similar toxicity (i.e., LC_{50} values at the same internal concentration) also provide support for the hypothesis that nonspecific hydrophobic toxicants cause toxicity principally by partitioning between aqueous and lipid phases. Therefore, the critical internal concentration may be similar for a wide range of hydrophobic agents.

While the data of Van Hoogen and Opperhuizen (1988) are striking with respect to their predictive utility for the three selected chlorobenzenes, Adams (1986) observed highly variable (i.e., 122-fold) tissue concentrations of 2,3,7,8-TCDD at the time of death in fathead minnows. The whole body residues at the time of death appeared to be a function of both exposure time (12–28 days) and dose (1.7, 6.7 and 63 ppT). The toxic response in this study occurred during the kinetic phase, since the time to 90% of steady state was calculated to be 48 days. These data reveal the need for caution when using models to extrapolate to agents of differing chemical class and of significantly differing physical-chemical and toxicological properties.

The proposed approach of McCarty (1986) has the *potential* to offer useful toxicological predictions of acute and chronic bioassays for chemicals within a similar grouping if information on potency and toxicokinetic parameters is provided. This approach may also be relevant for assessing the significance of organic toxicant concentrations observed in animals sampled in the field or other areas where the exposure regime is not known. He suggests that the body burden in an animal may be compared to body burdens previously determined to cause adverse effects (acute and chronic), and an estimate of risk and effect on the likelihood of survival of the organism or population may be possible. In the case of the three chlorobenzenes studied by Van Hoogen and Opperhuizen (1988), such an approach would be expected to yield values of direct relevance to site-specific ecological risk assessment. However, as noted above (Adams, 1986), this methodology has important limitations, especially when toxicity is a function of both exposure level and duration (e.g., as was observed for TCDD). In addition, biomonitoring information can be gathered in light of known kinetics, and a time-weighted average of the exposure history can be estimated. This type of data can be useful in determining whether observed toxicities are a function of the kinetic and/or steady-state phase of exposure.

The proposal of McCarty (1986) extended by Van Hoogen and Opperhuizen (1988) offers a method to link external dose with toxicological endpoints. However, the model predictions are based on the interrelationships of LC_{50} values, Kow and Kb. Within this context it is important to recognize that the relationship of Kb to Kow has some important restrictions. For example, large molecular weight agents are not efficiently transferred across the gill membranes, yielding a low Kb but with an often high Kow (Oliver and Niimi, 1985). In addition, Southworth et al. (1980, 1981) demonstrated that observed Kb values for polycyclic aromatic compounds metabolized by fish are much lower than predicted from their Kow. In such cases, the Kb process must be described by the addition of a metabolic elimination coefficient, further complicating the prediction and requiring additional data. For agents with long half-lives in fish, the use of laboratory BCF (kinetic or steady state) was shown to markedly underpredict field residues in the trout (Oliver and Niimi, 1985). The residues in the fish were considerably higher than could be explained from the bioconcentration from water. This represents a strong case that consumption of contaminated food was a major chemical source.

Oliver and Niimi (1985) argued that an equilibrium between fish chemical concentration and water chemical equilibrium is probably never achieved for such agents (e.g., DDT, DDE, PCBs). Thus, serious limitations exist using laboratory BCFs for predicting environmental concentrations in fish populations. In such circumstances, these authors recommended that a more preferred approach could be ''to measure the contaminant concentrations in the major food sources of the fish, to estimate rates of food consumption and to determine the chemical's half-life in the fish.''

Part II

Multimedia Ecological Risk Assessment Approaches

3

The Maximum Acceptable Tissue Concentration (MATissueC) Approach

A. CONCEPTUAL FRAMEWORK

Ecological risk assessment is concerned principally with site-specific ecosystem evaluation. Within this context, a wide range and variable number of species are present and their potential responses need to be assessed and/or estimated or predicted. As in the case of indicator chemicals, it is not feasible to identify and study each species for their responses to various contaminants so that acceptable dose rates or steady-state tissue concentrations can be determined.

A useful way to address this situation is to develop a properly calibrated and validated food web model that effectively estimates the transfer of contaminants through each trophic level. The role of the food web model in the ecological risk assessment process is that of a vehicle to assist in estimating exposure. The type of information that the food web model provides is an exposure estimation for each animal in the model. The model also has the capacity to estimate tissue levels in animals at known environmental exposure levels. The output data provided by the typical food web model are extremely useful to the ecological risk assessor. However, these values are in concentration ($\mu g/g$), while the usual exposure units in mammalian toxicology studies are in the form of a dose rate (mg/kg/day or mg/cm^2/day). Most toxicological studies do not report tissue levels of contaminant. Therefore, the tissue level concentrations are usually not available for reported exposure rates. However, these values (dosage rates and tissue concentrations) are interpredictable if toxicokinetic parameters, such as the rates of uptake and elimination, are known.

The food web model can be used to estimate how the tissue concentrations in each animal will change with different source terms (e.g., soil, sediment). Unfortunately, the model cannot directly address the biological/toxicological significance of the estimated tissue concentration of contaminant among the species of the food web. The key to a successful ecological risk assessment is making the link between food web information and toxicological predictions.

One approach to food web modeling for ecological risk assessment purposes has been developed by Fordham and Reagan (1991; HLA, 1991). The proposed food web model is employed to calculate the biomagnification factor (BMF) for contaminants of concern (COC) in both aquatic and terrestrial systems. The approach utilizes the BMF along with biota tissue concentration to estimate the source or medium concentration to which biota are exposed. When the tissue concentration value employed is measured in the environment, the BMF can be employed to estimate the source (medium) concentration in the environment. Taken one step further, when a maximum allowable *tissue* concentration (MATissueC) is employed, the BMF can be used to determine a maximum allowable source (medium) concentration or probabilistic biotic criterion (PBC) for that medium. For example,

$$\text{PBC (water)} = \frac{\text{MATissueC of a Generic Predictor (GP)}}{\text{BMF of the GP (water)}}$$

where PBC (water) is based only on an aquatic food chain. A similar scheme is used to derive a PBC for soils and sediment for aquatic and terrestrial food chains.

B. DERIVING MATissueC AND PBCs

The following is the recommended procedure for derivation of the MATissueC and its application for determining acceptable environmental medium concentrations.

1. Procedural Steps

1. Conduct a hazard assessment for the agent.
2. Determine each toxic endpoint caused in biological systems.
3. Derive a dose-response relationship for each endpoint of each agent.

4. Determine LOAELs and NOAELs* for each endpoint for each organ system.
5. Determine whether any extrapolation procedure is needed such as interspecies extrapolation, life stage extrapolation, etc., to derive an acceptable daily exposure rate (mg/kg/day).
6. At the acceptable exposure rate, conduct a study to estimate the concentration of toxicant in the target tissue. This tissue concentration should be at steady state and would be associated with a noninjurious level of exposure to the toxicant.
7. Once the MATissueC is derived for each species (or subgroup) of concern in the food chain (web), it will be necessary to determine the BMF. This may be accomplished by a detailed study of each species (or subgroup) of concern in the food web. The BMF of a terrestrial species has traditionally been estimated by measuring the concentration of the agent in the whole body or fat and dividing this value by the concentration in the food.
8. At this point the PBC can be calculated.

In theory, once a MATissueC has been derived, it would be necessary to develop a sampling strategy for each species of concern and assess their target organ tissue level. If the value was in excess of the MATissueC data, the source (medium) concentration would have to be decreased based on the above formula to achieve the acceptable level (MATissueC).

C. UNDER WHAT CIRCUMSTANCES SHOULD THE MATissueC METHODOLOGY BE USED?

The MATissueC methodology has its principal application for the species on which the data are collected. This presumes that the assumption of steady state is reasonably achieved. At issue, though, is whether the MATissueC approach can be employed to estimate acceptable MATissueC values for other species via an extrapolation process.

The measurement of tissue/whole body concentrations at a steady-state NOAEL is of considerable value. This represents a tissue concentration that has theoretically integrated all relevant pharmacokinetic processes and established a steady-state concentration at the site of action that is without harm. While there are numerous exceptions, especially in the case of receptor mediated toxicity

*See section at the end of this chapter for the recommended manner for defining the NOAEL.

(Portier et al., 1993) (e.g., dioxin, peroxisome proliferating agents), it is widely recognized that once pharmacokinetic processes are normalized (i.e., a similar steady-state concentration at the site of action), the vast majority of interspecies variation in susceptibility markedly diminish. Two species, at the same steady-state concentration, therefore, would be expected to behave in a similar fashion. For example, numerous studies have revealed vast interspecies differences in susceptibility to barbiturate-induced sleeping time, yet upon awakening (as measured by the righting reflex) all species studied displayed the same plasma concentrations (Brodie, 1962; Calabrese, 1991). This indicates that the receptor response (i.e., pharmacodynamic process) is similar among the species, while pharmacokinetic processes most likely account for the profound differences. However, once the NOAEL steady-state is achieved (e.g., MATissueC), then interspecies differences become profoundly reduced.

The dosage required to achieve a specific steady-state concentration may markedly vary among species based on variation in uptake (K1) and in elimination (K2) rates. A larger species with typically slower metabolism would achieve a given steady-state concentration at a lower rate of exposure than a smaller, more metabolically active species.

A genuine MATissueC at a steady-state NOAEL is likely to be a reasonable *first* approximation for a MATissueC for other species. The MATissueC approach requires that certain a priori criteria should be considered and satisfied prior to MATissueC implementation in ecological risk assessment.

1. Determination of whether the agent tissue/whole body level measurement can predict toxicity outcomes over a prolonged period of time. Unless this can be established the MATissueC approach cannot proceed.
2. Quantitative estimation of the extent to which steady-state has been achieved must be determined.
3. Knowledge of the principal basis for interspecies variation in susceptibility to the agent in question: if variation is principally pharmacokinetic then it is possible to proceed with the MATissueC approach if the steady-state assumption is acceptably satisfied.
4. If the principal basis for interspecies variation is pharmacodynamic it is less likely that interspecies toxicity responses will be comparable at similar steady-state tissue levels. This condition would place important limitations and/or greater uncertainty with the MATissueC approach.

5. Careful assessment of the health effects of exposure at the MATissueC needs to be made. This judgment should be guided to the extent possible with knowledge from the published literature on the health effects, including target organ and dose-response of the chemical in question.

The MATissueC approach is susceptible to error since tissue concentrations are available on only those animals who are survivors. Even when adverse effects may be encountered it will be difficult to ascribe cause-effect relationships because of the cross-sectional (i.e., snapshot) nature of this methodology as well as the possible presence of other agents that may co-vary with the chemical under study. These factors re-emphasize that unless a reliable toxicological database exists on the agent relating NOAEL/LOAEL values with tissue/whole body concentrations, the MATissueC approach could be of low reliability.

The MATissueC approach, therefore, is a tool for the ecological risk assessor. Its considerable attractiveness derives from its capacity to represent the integration of multimedia exposure with health effects in species of concern. Its attractiveness, though, must be tempered with the recognition of its limitations and the need to quantitatively factor in such restrictions in performing any ecological risk assessment.

1. Role of UFs in the MATissueC Approach

In this context, it is recognized that any predictive scheme will have uncertainties. This is recognized in human risk assessment processes with respect to RfD derivation and in low-dose cancer risk assessment processes. Both of these risk assessment methods are generally recognized as attempting to establish biologically plausible procedures within a regulatory context that errs on the side of safety. With the opportunity to employ the use of the MATissueC approach in ecological risk assessment comes the need to identify and quantify areas of uncertainty.

a. UF for the BMF

Even though the MATissueC integrates pharmacokinetic factors, it remains necessary to account for possible interspecies/intraspecies variation in pharmacokinetic parameters. This is accomplished by adopting the equivalent of an UF for the BMF. In this respect the BMF could receive an inter/intraspecies UF to account for differential rates of uptake and depuration. For example, assume species A has a NOAEL (mg/kg/day) associated with an MATissueC (e.g., whole body or organ concentration) of 10 mg/kg and a BMF of 0.1. The PBC would equal 10 mg/kg/0.1 = 100 units (ppm). If the PBC is needed to protect species B, but the NOAEL, MATissueC and BMF are not known, it is recommended that a BMF-UF be assigned based on phylogenetic relatedness following the recommendations for the interspecies UFs as given in Chapter 4, Table 4.14. Assume a species within genus comparison (i.e., BMF UF of 10-fold relative to species A). This would yield the following PBC: 10 mg/kg/1.00 = 10 units (ppm) such that the environmental concentration (PBC) would be some 10-fold lower than for species A. However, it is possible that species B may have a known BMF value but not a NOAEL with an MATissueC. If that were the case, a generic BMF UF would not be needed; the MATissueC of species A would be divided by the BMF of species B to estimate the species B PBC as follows: 10 mg/kg/.05 = 200 units (ppm).

Thus, one may be able to derive a tailored PBC if the BMF of the species of interest were known along with the MATissueC and the NOAEL of a relevant predictive species.

The concept of a BMF-UF is a reasonable one and represents an efficient, pharmacokinetically founded basis for the derivation of a PBC when interspecies uncertainty exists. As suggested above, the concept is also directly applicable to issues of intraspecies variation in susceptibility by using a generic or tailored UF for the BMF. A generic BMF-UF for intraspecies (life stage) extrapolation would be 10, since the individuals selected in the MATissueC approach are assumed to be more representative of the healthier subsegment of the population (e.g., similar to a healthy working population in human terms) (see Chapter 4, Section 9.3 for a further discussion of this issue.

In a similar manner, a BMF-UF of 10 for less-than-lifetime exposures (i.e., < 15% of normal lifespan) is adopted consistent with the rationale and recommendations for the TRV-process as presented in Chapter 4.

b. Protecting Endangered Species

Protecting individuals of an endangered species is comparable to the philosophy of protecting high risk groups seen in human risk assessment. In this case, there is the requirement to specifically incorporate an interindividual UF for noncarcinogens. If this is to be done in the context of an MATissueC approach it may require an additional or larger UF for intraspecies variation. Ideally, an additional UF for endangered species should be considered on a case-by-case basis since the size of the UF may be a function of the number of individuals in the population and the quality of toxicological data, among other factors. However, as an interim measure we would recommend a generic BMF UF for endangered species of 20, as recommended for the intraspecies UF for endangered species and 10-fold for a less-than-lifetime exposure, which is defined as <25% of normal lifespan consistent with the TRV process (Chapter 4, Table 4.14).

The case can also be made for considering cancer risk estimation for endangered species. This would, however, represent a radical departure from previous endpoint consideration in ecological risk assessment. Nonetheless, the intent of legislation to protect endangered species is to ensure that individuals are protected from environmentally induced harm. Thus, the use of quantitative risk assessment procedures could be employed for the assessment of cancer risks for endangered species. What level of risk would be acceptable (e.g., 10^{-3}, 10^{-4}, 10^{-5}, 10^{-6}) would be a policy decision.

c. UFs for LOAEL to NOAEL in the MATissueC Approach

If the only data available to derive an MATissueC were based on a LOAEL, this would create uncertainty with respect to what the true NOAEL is. If the pharmacokinetics of the agent were adequately defined, then it may be possible to derive a model-based estimate of the tissue (i.e., whole body) concentration at a rate of exposure consistent with the LOAEL to NOAEL UF. If pharmacokinetic data were unavailable or inadequate, an assumption of first-order kinetics may be assumed.

In this way it would be possible to utilize the MATissueC approach in the face of inadequate but possibly realistic exposure effect scenarios via the incorporation of an UF in the MATissueC approach. The size of the UF could again be either tailored to a specific species and chemical if sufficient relevant information were available (i.e., tailored UF), or generic in nature, as given in Table 4.14 as high to low dose extrapolation (i.e., LOAEL to NOAEL, FEL to NOAEL).

2. Factors to Consider Prior to Implementing the MATissueC Methodology: Summary and Recommendations

The principal consideration in the MATissueC methodology is whether the data are adequate to support its implementation (e.g., whether the chemical is a candidate for the MATissueC, whether steady-state has been established, whether pharmacokinetic factors predominate in affecting variability in susceptibility, whether knowledge of dose-response relationships are adequate, etc.).

The MATissueC methodology requires a thorough consideration of the following areas prior to its application in ecological risk assessment activities:

a. Determination of Whether an Agent Would Qualify for the MATissueC Method

1. Requirement that tissue and/or whole body levels of parent compound/or metabolite are highly predictive of toxicity effects.
2. Interindividual/interspecies variations in susceptibility are predominantly the result of pharmacokinetic processes.

b. Conduct of the MATissueC Methodology

1. Determination of sampling strategy:
   a. Which populations or subpopulations to sample (considerations to include randomization)
   b. Sample size
   c. Time framework of the sampling: days, weeks, months, seasonally, etc.
2. Development of protocol to estimate steady-state for tissue and whole body values.
3. Development of uncertainty estimates associated with the steady-state estimation.
4. Estimation of the BMF in the species of concern.
5. Development of criteria to assess biological effects:
   a. How is a reference group selected in order to determine if the MATissueC is a NOAEL, LOAEL, or other?
   b. What endpoints are to be measured? What is the impact on endpoint selection, including its variability on sample size?
   c. Assessment of statistical power to resolve:
      i. Type 1 error
      ii. Type 2 error
6. Assessment of co-presence of agents that may affect toxicity of agent in question.

D. AN IMPROVED METHOD FOR SELECTION OF THE NOAEL

The current methodology for assessing the NOAEL involves identifying the highest concentration administered that does not cause a statistically significant or biologically significant response to treatment as compared to the control group. The LOAEL represents the lowest dose that is found to cause a biologically and/or statistically significant response to treatment compared to the control group.

Such a definition of the NOAEL can be problematic since it evaluates the NOAEL only in reference to the control group. It does not consider the statistical relationship of the NOAEL to the LOAEL. More specifically, the NOAEL may be statistically different or indistinguishable from the LOAEL. Yet, the NOAEL is similar with respect to RfD derivation whether it is indistinguishable from the LOAEL or not. At a conceptual level one may ask the following questions:

1. Is a "NOAEL" truly a NOAEL if it is indistinguishable from the LOAEL in a statistical sense?
2. Should the NOAEL be defined only in reference to the control?
3. Should the NOAEL be defined in reference to both the control and LOAEL?

It would seem that a "NOAEL" that cannot be distinguished statistically from either the LOAEL or the control would not be satisfactorily classified as a NOAEL.

It is, therefore, proposed that a new NOAEL paradigm needs to be adopted in which the relationship of the NOAEL is assessed not only with respect to the control but also with the LOAEL. A new definition of a NOAEL is as follows:

A new term—the quasi-NOAEL—represents that dose which is not distinguishable from the control or LOAEL. The quasi-NOAEL may be utilized if there is no "true" NOAEL. However, the quasi-NOAEL, being intermediate between a true NOAEL and a LOAEL, argues for the use of an intermediate UF.

The following situations illustrate examples comparing the present and proposed approaches for assessing the NOAEL, where "N" represents the NOAEL, "L" represents the LOAEL, and "qN" represents the quasi NOAEL.

Situation #1:
<pre>
           N L
Situation #1: C 1 2 3
               ‾‾‾‾
</pre>

This represents a control and three doses (1, 2, and 3). The line underneath C and dose 1 indicates that they are not significantly different in a statistical sense. However, dose 1 and dose 2 are significantly different statistically. In this case the present and proposed approaches agree; dose 2 would be the LOAEL and dose 1 would be the NOAEL.

Situation #2:
Example 1: C 1 2 3

C and dose 1 do not differ in a statistically significant manner; doses 1 and 2 do not differ in a statistically significant manner; doses 2 and 3 do not differ in a statistically significant manner. C and dose 2 and C and dose 3 differ in a statistically significant manner.

Present   N L   Proposed   qN L
Approach: C 1 2 3   Approach: C 1 2 3

The traditional approach would conclude that dose 2 would be the LOAEL and dose 1 the NOAEL. However, this approach is rejected in the proposed approach because dose 1 is not distinguishable from dose 2. Dose 1 would be referred to as a quasi- (but not real) NOAEL. If a quasi-NOAEL exists it should receive an intermediate UF such as 0.5 of the LOAEL to NOAEL UF if there is no "true" NOAEL. Another example of Situation #2 is seen with Example 2 below.

Present   N L   Proposed   qN L
Approach: C 1 2 3 4   Approach: C 1 2 3 4

Situation #3: C 1 2 3 4

Doses 1 and 2 do not differ in a statistically significant manner from C. Doses 2 and 3 do not differ from each other in a statistically significant manner; doses 2 and 3 do not differ in a statistically significant manner.

Present	N L	Proposed	N qN L
Approach:	C 1 2 3 4	Approach:	C 1 2 3 4

The traditional approach would conclude that dose 3 would be the LOAEL since this dose is statistically significant from the control. However, in the proposed approach dose 2 is rejected as the NOAEL because it is statistically indistinguishable from dose 3. This would make dose 2 a quasi-NOAEL. Dose 1 would represent the true NOAEL since it is statistically significantly different from the LOAEL yet is indistinguishable from the control.

1. Implications for Study Design/Risk Assessment

An implication of the present analysis is that more stringent criteria for defining the NOAEL may lead to decisions rejecting a presently utilized NOAEL for some RfD/RfC-defining studies. This would result in either the adoption of a quasi-NOAEL or a new NOAEL (e.g., probably the next lower dose) for RfD/RfC derivation. The adoption of a new NOAEL at perhaps that next lower dosage will lower the RfD/RfC by that proportional decrease, while adoption of the quasi-NOAEL (in the absence of a "true" NOAEL) would reduce the RfD/RfC by a factor of 5.

Such an analysis emphasizes the need to derive experimentally based true NOAELs. To better define a true NOAEL will likely affect the number and size of dosage groups. Thus, improved study design along with enhanced statistical power will lead to a greater likelihood of defining a true NOAEL, thereby reducing the need of UFs in conjunction with LOAELs and quasi-NOAELs. An improved study design will be "rewarded" if it is capable of defining a true NOAEL.

4

Uncertainty Factors for Ecological Risk Assessment

A. INTRODUCTION

The use of uncertainty factors (UF) in human risk assessment is well known, widely recommended, and implemented at the federal and state level. The types of UFs employed in human risk assessment have traditionally included those dealing with uncertainty to: interspecies differences, interindividual (intraspecies) variation, less- than-lifetime (LL) exposures, and extrapolation from a dose that defined a lowest observable adverse effect level (LOAEL) to a no observable adverse effect level (NOAEL). In addition, for uncertainty not covered by this series of UFs, the Environmental Protection Agency (EPA) uses an additional UF factor called a modifying factor to address the residual uncertainty area(s). In Table 4.1, the EPA provides a description of each UF and its proposed magnitude.

No comparable articulation of the use of UFs in ecological risk assessment has been recommended by expert committees or advisory organizations such as the National Academy of Sciences or by federal regulatory agencies such as the EPA. Nonetheless, the use of UFs in ecological risk assessment has a long history, has been widely discussed, is not viewed as inherently controversial (ASTM 1978; Slooff et al. 1986), and is recommended for use under certain circumstances at hazardous waste sites such as the Rocky Mountain Arsenal (RMA) (HLA 1991). The use of the UF concept in ecological risk assessment has also been employed under a variety of descriptive terms such as application factor (AF) (Kenaga 1982) and assessment factors (EPA 1984).

Table 4.1.   Types of Uncertainty Factors (UFs) Used in Human Risk Assessment

Standard Uncertainty Factors (UFs)

Use a 10-fold factor when extrapolating from valid experimental results from studies using prolonged exposure to average healthy humans. This factor is intended to account for the variation in sensitivity among the members of the human population. (10H)

Use an additional 10-fold factor when extrapolating from valid results of long-term studies on experimental animals when results of studies of human exposure are not available or are inadequate. This factor is intended to account for the uncertainty in extrapolating animal data to the case of humans. (10A)

Use an additional 10-fold factor when extrapolating from less than chronic results on experimental animals when there are no useful long-term data. This factor is intended to account for the uncertainty in extrapolating from less than chronic NOAELs to chronic NOAELs. (10S)

Use an additional 10-fold factor when deriving a RfD from a LOAEL, instead of a NOAEL. This factor is intended to account for the uncertainty in extrapolating from LOAELs to NOAELs. (10L)

Modifying Factor (MF)

Use professional judgment to determine another uncertainty factor (MF) which is greater than zero and less than or equal to 10. The magnitude of the MF depends upon the professional assessment of scientific uncertainties of the study and database not explicitly treated above; e.g., the completeness of the overall database and the number of species tested. The default value for the MF is 1.

Source: Adapted from Dourson, M. L., and J. F. Stara, *Regulatory Toxicology and Pharmacology* 3:224–238 (1983).

This chapter will provide a direct comparison of how ecological and human risk assessments have incorporated the concept of UFs in their respective analyses. In addition, the chapter will provide the biological basis and toxicological rationale for deriving UFs for use in ecological risk assessment.

B. ACUTE-TO-CHRONIC EXTRAPOLATION UNCERTAINTY FACTOR

1. Introduction

Perhaps the most commonly employed and most readily accepted

UF in ecological risk assessment deals with acute to chronic extrapolation [i.e., the application factor (AF)]. This is based on the large experimentally derived acute toxicity database (i.e., 96 hour LC_{50}) for aquatic organisms and the need to derive chronic Maximum Allowable Toxicant Concentration (MATC) values.

The concept of the AF in the process of ecological risk assessment is employed in the prediction of chronic toxicity to organisms from known acute toxicity data within the same species. The AF was first proposed in 1967 by Mount and Stephan, environmental scientists specializing in aquatic toxicology for the U.S. EPA in Duluth, Minnesota. By definition, the AF is a ratio derived by dividing the 96 hour LC_{50} in an acute flow-through test into the no observed adverse effects exposure level (MATC) obtained in a chronic test for the same species. The experimentally derived ratio is then employed to estimate an MATC for other species or test conditions for which only acute LC_{50} data are available.

Mammalian risk assessment has generally not emphasized extrapolation from LD_{50}/LC_{50} values to chronic NOAELs (see McNamara, 1976 for an extensive review), although Layton et al. (1987) have proposed numerical schemes to estimate how extrapolation from acutely toxic doses to chronic NOAELS could be undertaken. Thus, the term AF as widely used in the field of aquatic toxicology has no history of use in the field of human risk assessment and is not mentioned in texts and articles in this area. Despite the general absence of discussion of how to extrapolate from acute to chronic values in human risk assessment, the range of recommended AFs (Kenaga 1978, 1982) are comparable to the observations of Layton et al. (1987) for human risk assessment when starting with LD_{50} data and estimating a chronic NOAEL (i.e., factor of 50–75-fold). The principal concern over the use of an UF for LD_{50}-type data has related to the probability that the acute response may be mechanistically unrelated to a chronic effect. It was thus believed that UFs for acute to chronic extrapolation are principally numerical values without adequate biological underpinnings. While this remains the prevailing view, it should be noted that Zeise et al. (1986) found a strong association between the LD_{50} and cancer potency in rodents. Such findings clearly support the need to reexamine the toxicological basis for acute to chronic relationships, and may provide a vehicle to derive a more biologically plausible rationale for the use of acute to chronic UFs as well as AFs.

2. Deriving Generic Application Factors Based on Acute to Chronic Ratios (ACR)

One approach to determine the ACR size has been to assess ratios of acute and chronic toxicity in aquatic organisms. Kenaga (1979) derived a large number of ACRs for pesticides and heavy metals based on bioassays with fish and daphnids. Acute to chronic ratios ranged over four orders of magnitude with such ACRs ranging from 1.1 to 11,100. A follow-up study by Kenaga (1982) assessed the chemical basis for why some agents have very large ACRs while others have ratios several orders of magnitude lower. In this subsequent analysis, data compiled from numerous sources utilized LC_{50} values from both static renewal and flow-through water systems and MATC values almost exclusively from flow-through water systems. Studies designed to derive MATC values employed both *partial* and *complete life-cycle* experiments. This assessment, based on data from 9 species of fish and 2 species of aquatic invertebrates, generated 135 AFs for 84 chemicals including chlorinated hydrocarbon insecticides (e.g., chlordane, heptachlor, endrin, DDT), fused ring aromatics (e.g., naphthalene), benzene and substituted benzenes (e.g., toluene), phenol and substituted phenols, halogenated aliphatics (e.g., trichloroethylene), herbicides (e.g., 2,4-D), cholinesterase-inhibiting insecticides (e.g., malathion), and miscellaneous organic and inorganic chemicals (e.g., cadmium, nickel, beryllium).

This analysis revealed a range of ACR values from 1 to 18,100. Approximately 86% of the chemicals displayed ACR values ≤ 100 regardless of which species was used to derive the ratio, and approximately 99% of all agents exhibited an ACR within a factor of 1,000 (Table 4.2). When Kenaga (1982) arranged the chemicals according to class, cholinesterase-inhibiting pesticides and heavy metals displayed the greatest percentage of ACR values above 125. Further analyses revealed no association between the magnitude of the ACR value and the degree of acute toxicity. For example, those agents with very low LC_{50} values did not display different ACRs than agents with moderate or low acute toxicity values. Likewise, this analysis did not reveal any predictive association between the magnitude of ACR values with parameters such as bioconcentration factor, persistence in the environment, or octanol-water partition coefficients.

Table 4.2.   Relationship of Vertebrate and Invertebrate Species to Acute-Chronic
Toxicity Ratio (ACR) Ranges for All Chemicals

ACR	All Species (%)	Fathead Minnow (%)	*Daphnia Magna* (%)
1–9	43.0	36	52.8
1–99	86.7	86	86.1
1–999	98.6	96	100.0
1–9,999	99.3	98	—
1–99,999	100.0	100	—
No. of examples	135.0	50	36

Source: Kenaga, 1982.

The AF as derived by Kenaga (1982) could be markedly affected, depending on the statistical approach employed to estimate the MATC as well as on the nature of the most sensitive endpoint. Refer to the discussion of Suter et al. (1987) in this chapter on "sensitive life-stage models" concerning statistical methods and endpoint selection and their impact on MATC derivation.

A practical application of acute-to-chronic UFs has been proposed by ESE (1989) with respect to their work at the Rocky Mountain Arsenal. Consistent with the above discussion that approximately 99% of all agents would be expected to be adequately handled by an UF of 1,000, the ESE report recommended the use of an acute LOAEL (i.e., LD_{50}) to a chronic NOEL UF of 1,000. This value reflected within species acute to chronic extrapolation and did not address interspecies variability.

3. Modeling Approaches

Slooff et al. (1986) attempted to predict chronic toxicity from acute lethality using correlation and regression analysis with fish and daphnia for 164 chemicals including pesticides, nonpesticides, organic and inorganic agents. A high correlation was shown between acute and chronic toxicity within a species ($r = .89$). The mathematical relationship between acute and chronic toxicity was determined to be $\log NOEC = -128 + 0.95 \log L(E)C_{50}$. Their findings are in strong quantitative agreement with the ACR methodology noted above for Kenaga (1979, 1982).

In contrast to the ACR approach for acute to chronic extrapolation, Suter et al. (1983a) used a least-squares regression model involving the natural logarithmic transformation of both dependent and independent variables to derive the MATC based on the best estimate of the regression. In this study, designed to provide an LC_{50} to MATC extrapolation, 45 data pairs from the literature, for which a life-cycle MATC and LC_{50} values were available, were evaluated for the same species from the same study. The equation for this extrapolation was determined to be:

$$\ln \text{GMATC (geometric MATC)} = 0.78 \ln LC_{50} - 1.87 \quad (4.1)$$

How the methodology of Suter et al. (1983a) would work is as follows. They provided an example in which it was desired to estimate the GMATC of the largemouth bass (*Micropterus salmoides*) for inorganic mercury based on the rainbow trout (*Salmo gairdneri*) LC_{50} of 249 $\mu g/L$. The largemouth bass belongs to the order Perciformes, while the rainbow trout belongs to the order Salmoniformes. Based on data from Johnson and Finley (1980) concerning some 503 pairs of Perciformes and Salmoniformes LC_{50} estimates, Suter et al. (1983a) developed a regression equation to estimate the LC_{50} in the largemouth bass (Figure 4.1) [Note: This is an interspecies (i.e. taxonomic) extrapolation which will be addressed in more detail elsewhere in this chapter.] The regression formula is:

$$\begin{array}{cc} \ln LC_{50} = 0.87 + 0.89 \ln LC_{50} & (4.2) \\ \text{(Perciformes)} \qquad\qquad \text{(Salmoniformes)} \end{array}$$

Based on this relationship, the point estimate of the LC_{50} for the largemouth bass for inorganic mercury is 324 $\mu g/L$. The authors then employed Equation 4.1 to estimate the GMATC for the largemouth bass (i.e., acute to chronic extrapolation), which turned out to be 14 $\mu g/L$. This, in effect, would yield an ACR value of 23.1.

An independent analysis estimating chronic values from LC_{50} values via regression modeling by Barnthouse et al. (1990) for aquatic species is generally consistent with the quantitative estimates of Kenaga (1982). The regression modeling may be used to derive an UF, depending on the desired level of protection (e.g., 95%, 99%). These regression-derived values are referred to here as 95% UF or 99% UFs. They may be defined as the *minimum ratio of the estimated toxicity value and its upper or lower prediction limits after back transformation.* In this case, 99% UFs are generally close to

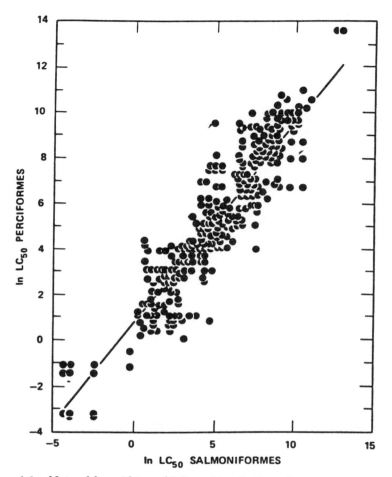

Figure 4.1. Natural logarithms of LC_{50} values for Perciformes plotted against Salmoniformes (orders of the same class, Osteichthyes). The solid line represents the least-squares linear regression of the natural logarithm of LC_{50} values for Perciformes species on the natural logarithm of LC_{50} values for Salmoniformes species. Data from Johnson and Finley (1980). *Source:* Suter et al., 1983a.

200-fold (i.e., weighted mean 264.9), with the most sensitive parameter (i.e., egg weight) having a 99% UF of 2,247 (Table 4.3).

The data from Table 4.3 indicate, therefore, that the size of the UF for acute to chronic extrapolation depends both on the endpoint measured and the degree of protection desired. Based on

Table 4.3.   Acute-Chronic Extrapolation: Means and Weighted Means Calculated for the 95% and 99% Prediction Intervals for Uncertainty Factors Calculated from Regression Models

			Uncertainty Factor		
			Prediction Interval		
X Variable	Y Variable	n	90%	95%	99%
Acute-Chronic Extrapolation					
LC_{50}	Hatch EC25	31	42	50	67
LC_{50}	Parent Mort EC25	28	27	32	43
LC_{50}	Larval Mort EC25	89	26	31	41
LC_{50}	Eggs EC25	42	53	63	84
LC_{50}[a]	Fecundity EC25	26	41	50	68
LC_{50}[a]	Weight[b] EC25	37	43	52	70
LC_{50}[a]	Weight/Egg EC25	14	200	245	344
Mean				74.7	102.4
Weighted Mean				54.3	73.5

[a]Regression analysis from Suter et al. 1987.
[b]Decrease in weight of fish at end of larval stage.

these data the 95% UF and the 99% UF would be reasonably approximated by 50 and 200, respectively, using weighted mean values for the following endpoints (e.g., hatch EC25, fecundity EC25).

It is important to consider the implications of whether the ACR UF was based on incomplete or complete life-stage data. In the case of Kenaga (1982), data on both incomplete/complete life-stages were employed. The data of Barnthouse et al. (1990) can be related to this life-stage dichotomy (i.e., incomplete vs complete) by consideration of specific endpoints (e.g., larval mortality, hatchability, etc.).

If the ACR UF were based on incomplete life cycle data it would appear that an additional UF addressing the incomplete life-stage is necessary. If the ACR UF were based on complete life cycle endpoints, then the sensitive life-stage UF should not be used. Of concern is how to select the endpoint upon which to base the regression. The large differential sensitivity and variability in endpoint response (e.g., hatch EC25 versus egg weight EC25) is problematic, and represents an area in need of further consideration.

If a frankly toxic effect other than lethality (e.g., LC_{50}) occurs, it is recommended that the UF be reduced to an intermediate position between the LOAEL to NOAEL UF (i.e., 10-fold, see below) and the acute toxicity value to NOAEL UF (i.e., 50-fold). This is consistent with dose response functions in toxicology and would hold for nonendangered and endangered species.

4. Comparison of Acute to Chronic Ratio Approach with Modeling

The question then arises as to whether there is a best or most appropriate way to estimate the MATC from acute data. Suter et al. (1983a) compared three approaches, including their regression analysis and two AF (i.e., ACR) methods. The AF methods were: (1) a GMATC derived directly with the fathead minnow (FM) for the chemical in question (i.e., FM-GMATC) and (2) a GMATC derived from the ratio of LC_{50} values for the species in question, based on the LC_{50} and GMATC values for the fathead minnow for the same chemical (i.e., AF-GMATC approach). Table 4.4 compares the three approaches to the true GMATC based on tested MATCs for each species and chemical. No approach distinguished itself, with each estimation procedure being closest to the GMATC in six cases and with the FM-GMATC and AF-GMATC estimates tying once. Both the FM-GMATC and AF-GMATC approaches had two instances where the error exceeded 10, while the extrapolation (i.e., regression) method had three such instances. The largest errors were found for the FM-GMATC derivation for malathion with blue-gill and flagfish where it underestimated toxicity by some 68- and 34-fold, respectively. The third highest error was reported for the regression technique where it overestimated toxicity by 25-fold for zinc in the brook trout. Of the 11 instances where the FM-GMATC method was in error by a factor of two or greater, eight underestimated the risk. Of the 11 instances where the AF-GMATC method was in error by a factor of two or greater, seven underestimated the risk. Of the 13 instances where the regression technique was in error by a factor of two or more, seven underestimated the risk.

This comparison revealed that all three methods provided estimates of the GMATC for another species that were generally within one order of magnitude of the true (i.e., experimentally derived) GMATC. In seven of the 54 (12.9%) comparisons, the estimates were in error by more than one order of magnitude but not greater than two orders of magnitude. In addition, in the true GMATC derivation experiments, the AFs ranged from .0002 to .8530 (4265-fold range) which is generally comparable to that reported earlier for Kenaga (1982). Within this context it is important to note that 11 of the 18 examples had ACR values within 99-fold, while 15 of

Table 4.4. Comparison of Methods for Estimating the Geometric Mean of the Maximum Acceptable Contaminant Concentration (GMATC)

Chemical	Species	True GMATC[a]	FM GMATC[b]	AF GMATC[c]	Extrapolation GMATC[d]
Atrazine	Bluegill	0.218	0.330	0.149	0.098 *
	Brook trout	0.088	0.330 *	0.109	0.075
Cadmium	Bluegill	0.050	0.046	0.135 *	0.253 *
	Flagfish	0.006	0.046 *	0.016 *	0.042 *
Chromium	Brook trout	0.265	1.987	3.257 **	0.584 *
	Rainbow trout	0.265	1.987 *	3.809 **	0.662 *
Copper	Bluegill	0.029	0.025	0.204 *	0.021
	Bluntnose minnow	0.009	0.025 *	0.017	0.005
	Brook trout	0.013	0.025	0.019	0.003 *
Lindane	Bluegill	0.011	0.015	0.006	0.001 **
	Brook trout	0.012	0.015	0.006	0.001 **
Malathion	Bluegill	0.005	0.341 **	0.004	0.003
	Flagfish	0.010	0.341 **	0.001 *	0.008
Methyl mercury	Brook trout	0.0005	0.0001 *	0.0001 *	0.0019 *
	Flagfish	0.0002	0.0001 *	0.0004 *	0.0055 *
Toxaphene	Channel catfish	0.0002	0.00004 *	0.0001 *	0.0005 *
Zinc	Brook trout	0.853	0.088	0.355 *	0.035 **
	Flagfish	0.102	0.088	0.266	0.027 *

Source: Suter et al., 1983a.

[a] True GMATC based on geometric mean of tested MATCs.

[b] FM GMATC based on GMATC for fathead minnow tested with same contaminant.

[c] AF GMATC based on application factor or ratio of LC_{50} values for listed species to LC_{50} values for fathead minnow for same contaminant.

[d] Extrapolation GMATC calculated by Equation 4.3. The best estimate of the true value is in italics; estimates which differ from the true GMATC by a factor of 2 or greater and by 10 or greater are indicated by single and double asterisks, respectively.

18 were within 999-fold. Three had ACR values of greater than 1,000.

While this direct comparison did not reveal a preferred approach, it must be emphasized that the comparison was limited to only nine agents in widely differing chemical classes and generalizations cannot be reliably made. However, a toxicologically-based evaluation can be used to differentiate the approaches as to their inherent capacity to yield biologically defensible predictions.

The FM-GMATC approach assumes that the fathead minnow would have identical LC_{50} and GMATC values as the species of interest. Since no data exist on the species of interest, the prediction relies entirely upon the capacity of the fathead minnow to simulate the response of the species of interest. This approach involves a direct interspecies extrapolation. It does not involve an acute to chronic extrapolation, since experimental data are collected for both types of endpoints.

In contrast to the FM-GMATC, the AF-GMATC utilizes experimental data from the fathead minnow on acute and chronic endpoints to derive an acute to chronic ratio. This ratio (not the absolute values) is believed to be identical with that of the species of interest. The key difference between the FM- and AF-GMATC methods is that the AF-GMATC method has acute data on the species of interest. This represents a substantial improvement over the FM-GMATC, since it should eliminate much of the uncertainty inherently present in the FM-GMATC approach. The uncertainty (i.e., interspecies) of the AF-GMATC approach is that the acute to chronic ratio of the fathead minnow is assumed to be identical to that of the species of concern. The AF-GMATC would be expected to be a better predictor than the FM-GMATC approach based on the extensive analyses of Slooff et al. (1986), which indicated that interspecies extrapolation was much more uncertain than estimates of acute to chronic ratios.

The strongest of the three approaches is expected to be the extrapolation E-GMATC method because it makes use of a robust database of acute to chronic ratios to derive its regression equation and then employs the acute toxicity value for the species of interest to estimate the E-GMATC.

The critical issue is how reliable the regression equation estimate of a GMATC would be, compared to the ratio offered by the fathead minnow. If the fathead minnow were an excellent predictor for the species of interest, then clearly it would be a potentially attractive option. However, since this is not likely to be inherently known, it is usually more attractive to use a broader database. This would likely lead to predictions that are not especially excellent, but not far off the mark.

In the cases of the AF- and E-GMATC, the type of extrapolation should be viewed as that involving acute to chronic (i.e., high to low dose extrapolation). In contrast to the FM-GMATC method, these methods do not include interspecies uncertainty.

The work of Slooff et al. (1986) indicates that considerably greater uncertainty exists with interspecies rather than acute to chronic extrapolation. This perspective also supports the conclusion noted above that the FM-GMATC is the least attractive methodology of the three reviewed.

5. Recommendation

This analysis revealed that a substantial database exists upon which to derive an acute to chronic UF for use in ecological risk assessment. Various approaches including the ACR method and statistical modeling provide comparable estimates of the acute to chronic UFs with respect to the proportion of the population protected. The ACR also appears to be quantitatively similar for aquatic as well as terrestrial animals. These collective findings support the recommendation that an acute to chronic UF of 50 be employed if the intent to is provide protection at the 95% level. Other-sized UFs could be estimated, depending on the level of protection desired and endpoint selected.

C. LOAEL TO NOAEL UF

1. Comparison Between Human and Ecological Approaches

The most common high to low dose extrapolation seen in mammalian risk assessment covers a much more modest dosage range than typically seen in aquatic risk assessment involving extrapolation from a known LOAEL dosage to an unknown dosage approximating the highest NOAEL. While considerable ecologically-based acute to chronic (i.e., application factor) extrapolation examples exist, there is little evidence (see HLA, 1991) that LOAEL (LOEC) to NOAEL (NOEC) extrapolation procedures have been implemented in ecological risk assessment. In the 1991 HLA report, which was directed toward both aquatic and terrestrial animals, the concept of LOAEL to NOAEL extrapolation was used in a manner qualitatively comparable to that seen in human risk assessment.

While it is unknown precisely why the concept of LOAEL to NOAEL extrapolation has not been used widely in ecological risk assessment, it is most likely related to the long history in aquatic toxicology of emphasizing the determination of principally acute effects. Thus, the chronic effects database has only more recently begun to become robust with respect to a wide range of chronic endpoints. As the field of aquatic toxicology continues to evolve in such a manner as to incorporate the types of endpoints measured in mammalian toxicology (e.g., reversible toxic responses), there will become greater pressure to adopt the use of the LOAEL to NOAEL UF. The usual aquatic study design involves a concurrent control and five treatment groups. This wide range of treatment groups provides an enhanced opportunity to estimate both the NOAEL and LOAEL. In such instances extrapolation uncertainties will be markedly reduced. Secondly, Suter et al. (1983a) have advocated the adoption of a geometric MATC (GMATC) which was defined by calculating the geometric mean of the NOEC and LOEC. In theory this may more closely estimate the actual highest NOEC than simple acceptance of the highest experimentally derived NOEC.

There is no reason for the dichotomy between ecological and human risk assessment goals to preclude the LOAEL to NOAEL UF. In fact, it is likely that the NOAEL could be estimated via regression analysis at an a priori response level, assuming the presence of an adequate database.

2. Recommendation

The derivation of the LOAEL to NOAEL UF in ecological risk assessment may be on the basis of a generic UF, as seen in the case of human risk assessment (see Table 4.1), regardless of whether the species of concern was nonendangered or protected. Another legitimate approach for deriving a NOAEL from a LOAEL may be via the use of regression modeling. The biological and statistical rationale for the use of modeling is presented in Section E1 of this chapter. The decision to use a generic UF or modeling approach should be made on a weight of evidence basis including consideration of study design, statistical analysis, endpoints measured, and biological relevance.

D. INTERSPECIES (TAXONOMIC) VARIATION UF

1. Introduction

The use of an interspecies extrapolation factor in ecological risk assessment has been discussed by a variety of authors (Suter et al., 1983a; Barnthouse et al., 1987, 1990; Slooff et al., 1986), and EPA (EPA, 1984, Stephan and Rogers, 1985). While all acknowledge that interspecies variation exists, no uniform approach has been presented to derive an UF to account for such variability. This lack of consensus of how to deal explicitly with interspecies variation for the purposes of ecological risk assessment stands in marked contrast to the generally accepted use of an interspecies UF of 10 in human risk assessment.

It may be argued that human risk assessment has had an inherently easier time since its goal (i.e., protecting humans) is more clearly defined and limited, while ecological risk assessors must consider interspecies differences for extrapolation purposes over a broad range of taxa with the goal of protecting not just one species but the ecosystem itself. Thus, the ecological risk assessor is confronted with a more formidable challenge so that a simple factor of 10 may be inadequate to deal with this apparently broader range of uncertainty. While extrapolation across numerous taxa may seem to be in the domain of the ecological risk assessors, how they "solve" the problem is likely to be of considerable theoretical and practical interest to the field of human risk assessment, since this could evolve procedures by which nonmammalian models could be used for extrapolation to humans.

2. Magnitude of Interspecies Variation

Numerous attempts have been made to assess the occurrence and magnitude of interspecies variation in response to ecological toxicants. Such studies have typically focused on interspecies variation with respect to acute toxicity in the aquatic environment (Pearson et al. 1979, Kenaga and Moolenaar 1979; Kenaga 1978, 1979; Suter et al. 1983a; Kimerle et al. 1983; Maki 1979; Doherty

1983; LeBlanc 1984; Slooff et al. 1986; Niederlehner et al. 1986). While these studies generally involved a wide range of agents with only a limited number of species, the comparative susceptibilities of aquatic species in general were shown to be contingent on their taxonomic relationship as well as the chemical tested. Table 4.5 shows the extreme range of EC_{50} values for cadmium as a function of taxonomic grouping (Niederlehner et al. 1986). In addition, while invertebrates and fresh and saltwater fish responded in a comparable manner to nonpesticide organics (r ranging from 0.79 to 0.95), no association was seen for pesticide susceptibilities between fish and invertebrates (r = .02)(LeBlanc 1984). Likewise, an assessment of acute toxicity by Suter et al. (1983) on 28 fish species from 17 genera, 10 families, and 6 orders for 271 chemicals (75% of which were pesticides) revealed a declining correlation (r) with increasing taxonomic distance (i.e., cogeneric species 0.90, genera 0.89, families 0.8, and orders 0.74) (Table 4.6).

Expansion of the database to include acute exposure values for 35 different species from 11 taxonomical groups to 15 agents was undertaken by Slooff et al. (1986). In their study, correlation and regression analyses were performed on log-transformed acute toxicity values for the 15 chemicals (Table 4.7) for each possible binary combination of species. The 95% uncertainty factors (UFs) provide an estimate of the variation in interspecies response (Figure 4.2).

Slooff et al. (1986) found a positive correlation in interspecies relationships in response to toxic agents with somewhat higher correlations being observed for species within the same phylogenetic grouping as compared to taxonomically more distant species. With respect to the 95% UF, despite the highly correlated relationship between species sensitivities to chemical, the UF values were quite variable and on occasion exceeded 1,000-fold. This extensive analysis resulted in an estimated 95% UF of nearly 1,200 for the various binary interspecies comparisons. Figure 4.3 displays the distribution of the interspecies UFs (Slooff et al. 1986). These results show that only a small proportion of the UFs were within a factor of 10, while the majority (60%) were between a factor of 10 and 100. A substantial percentage, approximately 30%, exceeded the 100-fold value, while about 11% exceeded 320-fold.

These findings led the authors to conclude that interspecies acute toxicity predictions possess greater uncertainty than predictions of chronic from acute effect levels in the same species (see previous

Table 4.5.   Summary of Single-Species Toxicity Test Data for Cadmium in $\mu g/L$

Taxonomic Family	Mean EC_{50}[a]	Mean EC_{95}[a]	Hardness Adjusted Mean EC_{50}[a] (at 65 mg/L)	Mean MATC[a]
Rotifers				
Philodinidae	311	458	597	—
Oligochaetes				
Naididae	1,700	—	2,304	—
Aeolosomatidae	2,445	6,115	1,448	43.28
Tubificidae	5,829	—	114,184	—
Lubriculidae	745	—	5,162	—
Turbellarians				
Planariidae	4,900	—	19,221	—
Mollusks				
Hydridae	1,600	—	970	—
Bithyniidae	8,400	—	11,383	—
Physidae	410	—	111	—
Planorbidae	201	665	201	—
Lymnaeidae	1,600	—	970	—
Cladocerans				
Daphnidae	39	93	52	1.39
Copepods				
Cyclopidae	340	—	1,334	—
Cyclopidae	—	15,650	—	—
Calanoididae	—	3,700	—	—
Ostracods				
Cypridopsidae	190	—	745	—
Isopods				
Talitridae	85	—	333	—
Amphipods				
Gammaridae	218	—	241	—
Insects				
Ephemerellidae	7,483	48,000	3,143	—
Heptageniidae	270	—	270	—
Pteronarcyidae	18,435	52,000	18,872	—
Odonate	8,100	—	10,977	—
Chironomidae	3,079	—	3,754	6.60
Culicidae	4,806	—	2,915	—
Trichopteran	3,400	—	4,207	—
Glossosomatidae	308,750	—	308,611	—
Hydropsychidae	5,750	—	5,747	—
Psephenidae	372,120	—	371,953	—
Bryozoans				
Pectinatellidae	700	1,364	190	—
Lophopodidae	150	3,550	41	—
Plumatellidae	1,090	3,508	296	—

continued

Table 4.5. *Continued*

Taxonomic Family	Mean EC_{50}[a]	Mean EC_{95}[a]	Hardness Adjusted Mean EC_{50}[a] (at 65 mg/L)	Mean MATC[a]
Fish				
Anguillidae	820	—	995	—
Salmonidae	3	—	7	3.60
Salmonidae	—	16,841	—	—
Cyprinidae	429	—	809	45.92
Cyprinodontidae	524	—	724	5.76
Poecillidae	2,445	17,488	6,569	—
Oryziatidae	213	—	87	—
Gasterosteidae	12,227	82,990	6,574	—
Percichthyidae	8,400	—	10,192	—
Centrarchidae	5,961	—	5,337	19.18
Esocidae	—	—	—	7.36
Catostomidae	—	—	—	7.10
Amphibians				
Ambystomidae	1,300	—	788	—
Pipidae	3,200	—	1,941	—

Source: Niederlehner et al., 1986.
[a]See text for definitions of terms.

section on application factors). The magnitude of the interspecies UF was reduced to some extent when comparisons were made within their same taxonomic group (e.g., bacteria, algae, protozoa, crustacea, insecta, pisces, amphibia, etc.). For instance, although no examples of regression derived UFs greater than 1,000 were seen for within taxa, comparison values greater than 100 were common.

The study of Slooff et al. (1986) is striking in its magnitude of interspecies comparisons. However, it is uncertain how the selection of species within taxonomic groups, as well as the number and range of chemical agents, affected the predictions. For example, one species was used to represent Coelenterata, Turbellaria, and Mollusca, seven species were employed to represent Insecta and Pisces, and the remaining taxonomic groups were intermediate in their species representation.

Slooff et al. (1986) provided no recommendations for how this information could or should be employed in the ecological risk assessment process. However, although not conclusive, the data of Slooff et al. (1986) provide a basis for assessing the association of

Table 4.6.   Listing of Acute Toxicity Comparisons at Four Taxonomic Levels (genus, family, order, class) for Data from Columbia National Fisheries Research Laboratory

Taxon I	Taxon 2	n	R^2
Species			
Salmo clarki	*Salmo gairdneri*	31	0.88
Salmo clarki	*Salmo salar*	7	0.67
Salmo clarki	*Salmo trutta*	7	0.93
Salmo gairdneri	*Salmo salar*	11	0.91
Salmo gairdneri	*Salmo trutta*	14	0.91
Salmo salar	*Salmo trutta*	5	0.99
Salvelinus fontinalis	*Salvelinus namaycush*	5	0.93
Ictalurus melas	*Ictalurus punctatus*	11	0.95
Lepomis cyanellus	*Lepomis macrochirus*	14	0.95
Genera			
Oncorhynchus	*Salmo*	54	0.93
Oncorhynchus	*Salvelinus*	18	0.81
Salmo	*Salvelinus*	85	0.83
Carassius	*Cyprinus*	6	0.96
Carassius	*Pimephales*	18	0.95
Cyprinus	*Pimephales*	8	0.92
Lepomis	*Micropterus*	48	0.92
Lepomis	*Pomoxis*	6	0.90
Families			
Salmonidae	*Esocidae*	9	0.18
Centrarchidae	*Percidae*	65	0.91
Orders			
Salmoniformes	*Cypriniformes*	246	0.76
Salmoniformes	*Siluriformes*	218	0.59
Salmoniformes	*Atheriniformes*	9	0.82
Salmoniformes	*Perciformes*	503	0.85
Cypriniformes	*Siluriformes*	98	0.80
Cypriniformes	*Atheriniformes*	5	0.99
Cypriniformes	*Perciformes*	218	0.73
Siluriformes	*Atheriniformes*	6	0.54
Siluriformes	*Perciformes*	204	0.59
Atheriniformes	*Perciformes*	11	0.94

Source: Suter et al., 1983a.

phylogenetic relatedness and the magnitude of uncertainty in interspecies extrapolation.

3. Interspecies Variation and Phylogenetic Relatedness

In a similar manner to the approach put forth by Slooff et al. (1986), we have estimated the 90% UF, 95% UF, and 99% UF of

Table 4.7. Agents Used in Toxicity Assays Reported by Slooff et al. 1986

Mercury(III)chloride
Cadmium nitrate
N-Propanol
n-Heptanol
Ethyl acetate
Ethyl propionate
Acetone
Trichloroethylene
Benzene
Aniline
Allylamine
Pyridine
o-Cresol
Salicylaldehyde
Pentachlorophenol

42 taxonomic binary toxicity comparisons of fish species published by Barnthouse et al. (1990). These assessments represent a quantitative estimate of interspecies toxicological predictions based on phylogenetic relatedness as seen at the species, genus, family, order, and class levels of organization. The data indicate that the extent of taxonomic variation is similar for the "species within genus" and the "genera within family" categories with values of approximately 6-fold at the 95% level, and 10-fold at the 99% level. However, the phylogenetic relatedness diminishes when one considers "families within order" and "orders within class," with the size of the UFs increasing appreciably. For example, excluding #46, the mean UF for "orders within class" was 23.5-fold for 95% and 31.7-fold for 99% levels, respectively.

The magnitude of the interspecies variation seen in this further evaluation of the Barnthouse et al. (1990) data support the premise that interspecies variation is generally inversely associated with phylogenetic relatedness. The magnitude of the 99% UF for "species within genera" was about 10-fold, while up to 32-fold for "orders within class." These findings are generally consistent with the above data of Slooff et al. (1986), although the absolute magnitude of interspecies variation is somewhat less in the Barnthouse et al. (1990) data. This is probably due to the fact that some of the Slooff et al. (1986) comparisons were across major taxa ranging from bacteria to amphibia.

The present analysis of the Barnthouse et al. (1990) data is restricted to the use of aquatic models. Whether similar relationships

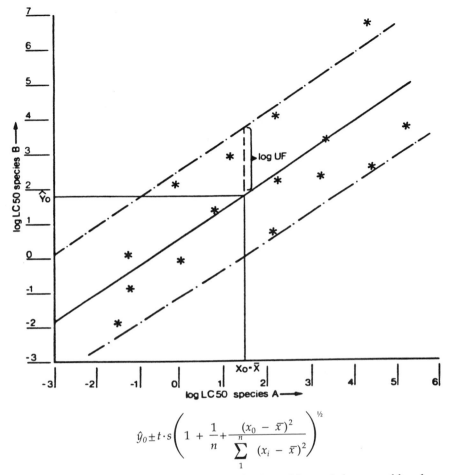

$$\hat{y}_0 \pm t \cdot s \left(1 + \frac{1}{n} + \frac{(x_0 - \bar{x})^2}{\displaystyle\sum_{1}^{n} (x_i - \bar{x})^2} \right)^{\!\!\frac{1}{2}}$$

where $t = 1/2$ percentile of a Student distribution with $n - 2$ degrees of freedom; s = estimated residual variance; n = number of observations; x = known log LC_{50} species A; and y = estimated log LC_{50} species B. When $x_0 = \bar{x}$, the prediction interval becomes $\hat{y}_0 \pm t \cdot s[1 + (1/n)]^{\frac{1}{2}}$. The uncertainty factor is defined as the minimum ratio of the estimated toxicity value and its 95% upper or lower prediction limit after back transformation: $UF = 10^{t \cdot s}[1 + (1/n)]^{\frac{1}{2}}$. In applied terms: If the toxicity of a given compound for A is known, the value for B is in the range of $A/UF < B < A \cdot UF$ with a probability of 95%.

Figure 4.2.   Determination of uncertainty factors (UF). Toxicity values were logarithmically transformed and the line of best fit was constructed through least-squares estimation. *Source:* Slooff et al., 1986.

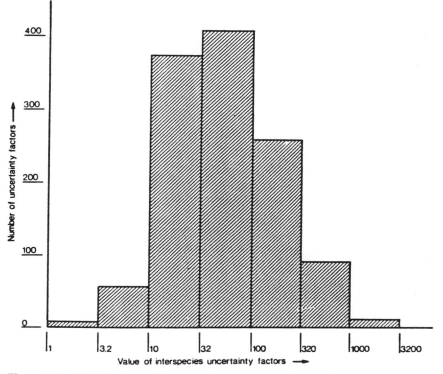

Figure 4.3. Distribution of interspecies uncertainty factors based on acute toxicity data for 15 chemical compounds and 35 freshwater species. *Source:* Slooff et al. 1986.

hold for terrestrial species is unknown. Nonetheless, the data analysis has the theoretical potential to offer a toxicologically based process for deriving UFs for ecological risk assessments based on phylogenetic relatedness. However, one must recognize that extreme outliers exist, such as #34 and especially #46, that must temper enthusiasm for the derivation of generic UFs for ecological risk assessment procedures. Despite these exceptions, the above approach offers an encouraging foundation upon which UFs could be derived and for which underlying biological regularities could be discerned for predicting the basis of interspecies variation.

4. Deriving an Interspecies UF for Ecological Risk Assessment: Recommendations

When deriving a generic interspecies UF for use in ecological risk assessment, it will be necessary to consider (1) the magnitude of protection built into the UF (e.g., 95%, 99%, or other), (2) the relevance of the chemicals comprising the model, (3) whether interspecies comparisons based on phylogenetic relatedness can be legitimately made to the chemicals of concern for a site-specific analysis, and (4) whether and how to develop a "weighted" UF (within each UF category) based on the available data (Table 4.8).

This exercise in UF estimation must be placed within the context of its limitations. For example, the approach used for UF derivation assumes that the species have been randomly selected from the broader population and are representative of that population. This fundamental assumption is not satisfied for the data of Table 4.8. This is not a minor consideration, but one that can drastically alter estimated values. However, the amount of data available for each category is also highly variable, with 102 comparisons offered for the species within genera, 212 comparisons offered for the genera within families, 125 comparisons offered for the families within orders, and 1803 comparisons offered for the orders within classes. The "orders within classes" comprises the vast majority of the database (80.4%). On this basis, it would appear that the most reliable database would be the orders within classes.

What should the interspecies UF be for ecological risk assessment? The data from Barnthouse et al. (1990) support the premise that the size should be a function of the degree of phylogenetic relatedness. However, to make a fair assessment, the categorical comparisons (i.e., species within genera, genera within families, etc.) need to be made on the same compounds under similar testing protocol. As it currently stands, the possibility exists that the reason for the phylogenetic relationship may be simply a matter of chemical selection and not interspecies variation. Until the proper comparison is made, the data from Barnthouse et al. (1990) present the best argument for interspecies UFs being based on phylogenetic relatedness.

Assuming that the individual binary comparisons are appropriate, we contend that the average of the upper percentiles (95%,

Table 4.8. Taxonomic Extrapolation: Means and Weighted Means Calculated for the 95% and 99% Confidence Intervals for Uncertainty Factors Calculated from Regression Models

X Variable	Y Variable	n	Uncertainty Factor Prediction Interval 90%	95%	99%
Taxonomic Extrapolation: Species within Genera					
Salmo clarkii	S. gairdneri	18	8	9	13
Salmo clarkii	S. salar	6	5	6	10
Salmo clarkii	S. trutta	8	4	6	8
Salmo gairdneri	S. salar	10	6	7	11
Salmo gairdneri	S. trutta	15	3	4	5
Salmo salar	S. trutta	7	4	5	8
Ictalurus melas	I. punctatus	12	4	5	7
Lepomis cyanellus	L. macrochirus	14	5	6	9
F. heteroclitus	Fundulus majalis	12	5	6	8
Mean				6.1	10.1
Weighted Mean				6.0	7.7
Taxonomic Extrapolation: Genera within Families					
Oncorynchus	Salmo	56	4	5	6
Oncorynchus	Salvelinus	13	3	4	5
Salmo	Salvelinus	56	5	5	7
Carassius	Cyprinus	8	3	4	6
Carassius	Pimephales	19	5	7	9
Cyprinus	Pimephales	10	5	7	10
Lepomis	Micropterus	30	7	8	11
Lepomis	Pomoxis	8	7	9	13
Cyprinodon	Fundulus	12	5	6	8
Mean				6.1	8.3
Weighted Mean				5.8	7.7
Taxonomic Extrapolation: Families within Orders					
Centrarchidae	Percidae	47	9	10	14
Centrarchidae	Cichlidae	6	3	4	6
Percidae	Cichlidae	5	10	13	24
Salmonidae	Esocidae	11	7	9	13
Atherinidae	Cyprinodontidae	32	6	7	9
Mugilidae	Labridae	12	45	55	78
Cyprinodontidae	Poecillidae	12	3	3	5
Mean				14.4	21.3
Weighted Mean				12.6	17.9

continued

Table 4.8. *Continued*

X Variable	Y Variable	n	Uncertainty Factor Prediction Interval 90%	95%	99%
Taxonomic Extrapolation: Orders within Classes					
Salmoniformes	Cypriniformes	225	17	20	27
Salmoniformes	Siluriformes	203	33	39	51
Salmoniformes	Perciformes	443	10	12	16
Cypriniformes	Siluriformes	111	9	11	15
Cypriniformes	Perciformes	219	27	32	43
Siluriformes	Perciformes	190	53	63	83
Anguiliformes	Tetraodontiformes	12	10	13	18
Anguiliformes	Perciformes	34	21	25	34
Anguiliformes	Gasterosteiformes	8	13	16	24
Anguiliformes	Atheriniformes	46	7	9	12
Atheriniformes	Cypriniformes	7	393	501[a]	786[a]
Atheriniformes	Tetraodontiformes	46	11	13	17
Atheriniformes	Perciformes	148	21	25	33
Atheriniformes	Gasterosteiformes	36	17	20	27
Gasterosteiformes	Tetraodontiformes	8	16	20	30
Gasterosteiformes	Perciformes	33	26	32	43
Perciformes	Tetraodontiformes	34	21	25	34
Mean				23.5	31.7
Weighted Mean				26.0	34.5

Source: Barnthouse et al., 1990.
[a]Not included in calculations.

99%) of experimentally derived values (i.e., Tables 4.6 to 4.8) not be used to derive directly the categorical interspecies UFs. *However, those values should be considered as representative of the larger and as yet untested population of potential binary comparisons within their respective categories.* With these values as input data, we propose to follow the scheme of Van Straalen and Denneman (1989) to estimate the upper 95% UF of the listing of 95% or 99% UF factors, assuming a logistic distribution of UFs within the category (Table 4.9). For the sake of argument, this would yield a UF of 10 for the species within genera category and up to 65-fold for the orders within class for the given individual 95% UF values. If the given individual upper 99% UFs were used, this would yield a 99% UF of 16 for the species within genera category and up to 88 for the orders within classes category. These data suggest that interspecies UFs for ecological risk assessment could range from a low of 10-fold for the species within genera to a high of roughly 100-fold for the orders within class category (i.e., 95% P.I.).

Table 4.9.   Upper 95% UFs Calculated for the 95% and 99% Prediction Intervals
Based on the Scheme of Van Straalen and Denneman (1989)

Regression Model	Prediction Interval	
	95%	99%
Species within genus extrapolation	10.0	16.3
Genera within family extrapolation	11.7	16.9
Families within order extrapolation	99.5	145.0
Orders within class extrapolation	64.8	87.5

The intermediate category UFs (i.e., genera within families, families within orders) could be assigned separate values intermediately spaced, such as 30 and 60, respectively. Table 4.10 summarizes the recommended UFs and how these UFs based on phylogenetic relatedness could be applied.

In the case of interspecies comparisons at the classes within phylum category, the most relevant data are from Slooff et al. (1986). Their findings, as discussed previously, support an approximate 95% generic UF of 1000-fold (Table 4.10).

E. INTRASPECIES UF

1. Sensitive Life-Stage UFs in Species Not Specially Protected by Legislation (e.g., Endangered Species)

a. Introduction

Ecological risk assessments frequently address the issue of life stage extrapolation. This situation exists when investigators have opted to perform experimentation on a presumed sensitive early life stage (e.g., hatching, larval survival, etc.) and then estimate a MATC for the adult. This type of extrapolation is seen to bridge a gap between both the traditional UF in human risk assessment for less-than-lifetime (LL) exposure and interindividual variation. The life stage extrapolation feature of ecological risk assessment occupies a legitimate place in both of these human risk assessment UFs. This is because the studies with young animals constitute a less-than-lifetime study duration. The experiments also point out that differential susceptibility in the population may exist as a function of age. The presence of the life stage extrapolation information

Table 4.10. Interspecies UFs for Ecological Risk Assessment: Listing and Application

Interspecies (Species within Genus)	UF	10
Interspecies (Genus within Family)	UF	30
Interspecies (Families within Order)	UF	60
Interspecies (Orders within Class)	UF	100
Interspecies (Classes within Phylum)	UF	1000

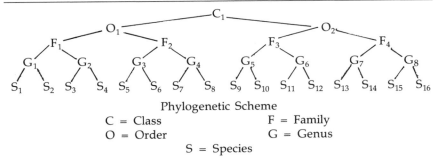

Phylogenetic Scheme

C = Class   F = Family
O = Order   G = Genus
        S = Species

Application of Interspecies UF Based on Phylogenetic Relatedness

	UF Size	Explanation for UF Selection
S_1 (data available) and S_2 (species of concern)	10	Species within Genus
S_1 (data available) and S_3 (species of concern)	30	Genera within Family
S_1 (data available) and S_5 (species of concern)	60	Families within Order
S_1 (data available) and S_9 (species of concern)	100	Orders within Class

in both of these human UFs indicates that these two UFs are not fully independent as generally assumed, but are to some extent interdependent.

The life stage UF does not address the remaining aspects of interindividual variation (e.g., sex differences, genetic variation, nutritional status, preexisting disease contributions). This is most likely because the age factor, being a developmental process, is more fundamental to species success, while the other factors are more limited in their range of influence on species survival. This also appears to be the case for the less-than-lifetime UF. The life stage factor is not concerned with differential susceptibility of the old, nor with the issue of cumulative damage as long as the reproductive success is assumed. This limited use of aspects of the interindividual UF and less-than-lifetime UF as seen in the life stage extrapolation scheme is clearly a function of the historical role of ecological risk assessment being concerned with the survival of the population and not the individual.

Long-term chronic experiments are life cycle studies that are designed to determine a lowest observed effect concentration (LOEC) and a no observed effect concentration (NOEC) on survival, growth and reproduction. Based on the results of such chronic studies an estimate of the MATC can be made (ASTM 1978). Fish chronic toxicity tests have traditionally been long-term exposures of up to 12 months in duration. However, due to the need for more rapidly required information and cost containment, much alteration has been directed to (1) aquatic species such as Daphnia where chronic toxicity tests can be conducted in 28 days, as well as to (2) fish species where the focus is on only a partial life-cycle test that includes the most sensitive life-stages (i.e., usually early development periods).

b. Sensitive Life-Stage Models

It has been argued that the estimation of MATC values from early life stage tests using criteria such as hatchability, survival, growth, and deformities was within a factor of 2 in most instances of the MATCs derived from chronic studies (McKim 1977; Macek and Sleight 1977). These findings were supported by Woltering (1984) who, based on data from 173 long-term fish toxicity tests, concluded that fry survival data would have estimated MATC values within a factor of 2 to 7.

Despite the above supportive argument in favor of the "short-term chronic" studies to estimate MATC values, Suter et al. (1987) have challenged its use. Based on an analysis of 176 tests on 93 chemicals with 18 species, the authors concluded that the measured endpoints (i.e., hatching success, larval survival, etc.) were consistently less sensitive than reduction in fecundity. According to the authors, the principal reason why fecundity has been overlooked as the most sensitive response is that prior statistical analyses were based on hypothesis testing and not on the levels of effect as estimated via regression analyses. More specifically, Suter et al. (1987) indicate that since fecundity is quite variable and only small numbers of fish are employed in this aspect of life-cycle tests, large reductions in fecundity may not be shown to be statistically significant. They rely heavily on the arguments of Stephan and Rogers (1985) that hypothesis-testing to define chronic effect benchmarks have important undesirable features since some MATCs have been set at levels associated with greater than 50% mortality.

This debate over whether to use regression techniques or hypothesis testing to determine endpoints for MATCs is of considerable practical importance. Stephan and Rogers (1985) argue that there are numerous computational and conceptual advantages using regression analysis over hypothesis testing for calculating results of chronic toxicity tests with aquatic animals (Table 4.11). A critical factor here is that most toxicity tests with aquatic animals are highly suited for regression analysis since they typically involve six treatments, including a control and five concentrations. This is in marked contrast to most mammalian studies which involve a control and only two treatments. It should be emphasized that in 1984 Crump proposed a comparable scheme for determining allowable daily intakes for humans based on mammalian toxicology studies.

In addition to statistical considerations, Suter et al. (1987) argue that even though the MATC is believed to be the threshold for fish populations, no thresholds or even negligible effects for most of the published chronic tests have been provided. Even the least sensitive endpoint, which was hatching success, displayed a 12% average reduction at the MATC, as determined by regression analysis. The most sensitive overall response (i.e., fecundity) displayed a 42% reduction at the MATC, as determined by regression analysis. These treatment effects, judged significant by regression analysis, were not deemed statistically significant via hypothesis testing.

The arguments concerning both sensitive endpoint and statistical method selection raised by Suter et al. (1987) are substantial both theoretically and practically with respect to how ecological toxicology studies are designed, conducted, and interpreted. The information also has a considerable relevance for multiple extrapolation issues, including acute to chronic UF, LOAEL to NOAEL UF, interspecies UF, and less-than-lifetime UF, since the use of the regression technique provides a direct estimate of the 95% UF discussed above.

c. Recommendation for Sensitive Life-Stage UF

The size of the sensitive life-stage UF which should be seen as a component of an overall intraspecies UF for use in ecological risk assessment may be argued as being in the range of 2- to 7-fold (McKim, 1977). However, the recent findings of Suter et al. (1987) concerning sensitive endpoint identification argue for a

Table 4.11.   Advantages for Using Regression Analysis to Estimate MATCs

Computational Advantages

1. Regression analysis provides a well-defined procedure for interpolation of effect to untested concentrations, whereas hypothesis testing provides quantitative information concerning only the concentrations that were actually tested.

2. Estimates of toxicity calculated using hypothesis testing are sensitive to the care with which the test was conducted and to the number of replicates used, whereas estimates of toxicity calculated using regression analysis are not.

3. The choice of ''alpha'' does not affect the estimate of toxicity obtained using regression analysis.

4. The estimate of the endpoint concentration obtained using regression analysis is independent of the concentrations actually used in the test.

5. Changes in the statistical procedure can affect the results of hypothesis testing more than comparable changes can affect the results of regression analysis.

6. Regression analysis can accommodate unexpected inversions in the data.

7. Regression analysis does not require treating experimental units as replicates if, in fact, they are not.

Conceptual Advantages

1. Use of regression analysis will encourage aquatic toxicologists to consider the real-world importance of the observed effects.

2. It is easier for toxicologists and others to make decisions about the adequacy of a toxicity test in terms of confidence limits on endpoint concentrations than in terms of a minimum statistically significant difference.

3. Use of regression analysis will encourage aquatic toxicologists to think of chronic toxicity in terms of a concentration-effect relationship.

4. Use of regression analysis will discourage people from thinking that hypothesis testing identifies ''no effect'' concentrations.

5. Use of regression analysis will discourage people from thinking that the most sensitive hypothesis test is the best hypothesis test.

Source: Stephan and Rogers, 1985.

somewhat larger UF. Support for this position is found in Table 4.12 (Barnthouse et al., 1990), which indicates that the 95% UF for Larval Mortality EC25, Hatch EC25, and EGG EC25 when compared to parent mortality EC25 range from 8-to 16-fold. These findings, while qualitatively similar to the McKim (1977) findings, support a rounded value of 10-fold for the sensitive life stage 95% UF.

Table 4.12. Life-Stage Extrapolation: Means and Weighted Means Calculated for the 95% and 99% Prediction Intervals for Uncertainty Factors Calculated from Regression Models

| | | | Uncertainty Factor | | |
| | | | Prediction Interval | | |
X Variable	Y Variable	n	90%	95%	99%
Life-Stage Extrapolation					
Larval Mort EC25	Parent Mort EC25	16	6	8	11
Larval Mort EC25	Hatch[a] EC25	28	11	13	17
Hatch EC25	Parent Mort EC25	7	10	13	20
Hatch EC25	Eggs[b] EC25	6	10	13	21
Eggs EC25	Parent Mort EC25	27	13	16	21
Eggs EC25	Larval Mort EC25	25	6	7	9
Mean				11.7	16.5
Weighted Mean				11.6	15.7

Source: Barnthouse et al., 1990.
[a]Fraction of eggs failing to produce normal larvae.
[b]Number of eggs produced per female fish surviving to the beginning of spawning.

2. Intraspecies UF for Protected Species

a. Background

While the pervasive approach inherent in ecological risk assessment is that of protection of the population (in contrast to the individual), this perspective is altered markedly when consideration is given to species protected by law under the Endangered Species Act and the Migratory Bird Treaty Act. In the case of specially protected species, the environmental legislation has the intention of ensuring protection for individuals as well as the population. Thus, the growth, maintenance, and reproduction triad, as previously discussed, may not be fully adequate to ensure protection of such species. In such cases it may be necessary to address additional toxic endpoints not usually considered in ecological risk assessment, such as chronic toxicities and interindividual variation in susceptibility. It may also be necessary to incorporate the use of larger UFs in the derivation of acceptable exposures for such species. This area is clearly one in which additional toxicologically based risk assessment criteria are needed for development.

While no specific guidance exists on UFs for endangered species, it bears considerable similarity to the human intraspecies UF where the goal is to protect susceptible subgroups within the population as well as developmental and reproductive domains. There are some relevant distinctions to be made, however, between the use of the intraspecies (i.e., interindividual) UF for humans and possible UFs for endangered species. First, the ostensible goal of the human intraspecies UF is to protect susceptible, yet reasonably sizeable, "subgroups" above a certain proportion in the population (e.g., 1%) while the Endangered Species Act is specifically concerned with protecting all individuals. This is interpreted to mean that uniquely sensitive humans or sensitive humans in small subgroups (e.g., 1% of the population) may not be specifically intended for protection. Second, the size of the distribution of humans is huge, while that of an endangered species would be expected to be quite small (dozens to several hundred). The degree of interindividual variation would be expected to increase as the population itself increases. For example, the variability could be expected to be larger in a population of 200 individuals as compared to a population of 20 individuals. A larger sized UF would be derived for the larger population if the goal were to protect *all* individuals. On the other hand, if only 20 individuals exist there is considerably less room for error of losing the entire species! Thus, since the stakes are so high (at least in a legislative sense) it is reasonable to err on the side of safety in such cases. This would tend to be the case for all endangered species evaluations, but it could be more pronounced in cases such as the 20 individuals example. Perhaps this is a situation (i.e., where the population is quite low, such as 20) where a modifying factor would be appropriate.

Should the size of the intraspecies UF for ecological risk assessments dealing with endangered species be the same as used in human risk assessment? It is argued here that the UF for Endangered Species should be larger than the 10-fold factor used in human risk assessment. This is based on the goal that each individual needs to be protected, and that the 10-fold factor may not address the full range of human interindividual variation in response to toxic agents, even when only relatively small sample sizes (i.e., several hundred) are used. It is highly uncertain what type of variation to expect for Endangered Species and it is likely to be considerably different, depending on the specific species.

Should each intraspecies UF for a protected species be the same? Ideally, it would be expected that protected species should have their own uniquely tailored UF; however, there is no obvious means to determine how to derive such a value. Unfortunately, it is highly likely that no toxicological information will be available for many Endangered Species.

b. Recommendation

It would appear reasonable for each protected species to have the same intraspecies UF due to a lack of species-specific data in this area. While the magnitude of this UF would be expected to be greater than the 10-fold factor used in the human risk assessment, how much greater? In the absence of an adequate database, it would appear that the present UF of 10-fold factor should be increased by a factor of 2, making it 20-fold. This is sufficiently large as to provide an apparent additional margin of safety, while at the same time not being unnecessarily conservative. It is a tentative judgment that needs to be subsequently reevaluated so that adjustments could be made.

F. LESS THAN LIFETIME (LL) UF

1. Nonendangered Species: Rationale and Recommendation

The use of a LL-UF is standard practice in human risk assessment processes; however, the concept of a chronic bioassay and its utility in ecological risk assessment theory and practice concentrates on the species rather than the individual. It is proposed that a "chronic" study in ecological risk assessment for nonendangered species be 15% of the normal adult life-span after weaning. This duration was selected since it would provide an adequate opportunity for reproductive success by permitting the animal the opportunity to achieve reproductive maturity. [When a species becomes reproductively mature, relative to their adult life-span, is variable according to the species and breed.]

Some would argue that the LL-UF for nonendangered species is already incorporated within the sensitive lifestage (i.e., intraspecies) UF and is unnecessary. We believe this argument is not compelling, since fecundity* is often the most sensitive endpoint for MATC derivation (Suter et al., 1987), and that it is uncertain in most

---

*Fecundity is defined by Suter et al. (1987) as viable eggs produced per female surviving to the initiation of reproduction.

instances whether these effects are principally the result of debilitation of the adults or to direct effects on reproductive process such as oocyte development.

Within the percentage of life-span context, a LL-UF of >1 would not be necessary if a study of ≥15% of a normal adult life-span after weaning has been achieved. This would essentially conform to a 90-day study in rodents and up to a 9–12 month study in a longer lived species such as the dog. These studies are designed to ensure that growth, maintenance, and reproductive functions would be sustained. Studies of a duration <15% in rodents would likely be an acute toxicity assessment, involving the derivation of an LD_{50}. This study would be best handled within the context of a frank effect level (FEL) to NOAEL UF extrapolation. Studies of <15% in dogs, such as those of 3 or 6 months, would not likely be of an acute toxicity nature. Such studies would need to be handled within the context of a LL-UF. The size of the LL-UF in ecological risk assessment is proposed to be 10 and would be consistent with that used in human risk assessment.

2. Endangered Species: Rationale and Recommendation

The concept of a LL UF changes when endangered species are considered. In this case, one is guided by the premise that not only is reproductive success important, but also the health of individual animals. Based on the current dictum that individuals of endangered species need to be protected, it is recommended that procedures by which a LL UF are derived in human risk assessment be adopted for endangered species. Guidance that exists on this issue is as follows:

1. Rodent studies of a typical ''experimental'' lifetime are for two years, which is about 50–70% of the expected average life-span of the mouse/rat, depending on the strain.
2. Generally, if the duration of a rodent study is less than a year, a case can be made for the use of a LL UF. The case has historically been much less compelling if the duration of the study is ≥1 year, but <2 years. This would essentially indicate that in mammalian toxicology if the duration of the rodent study were 25–35% of the expected average life-span for the studies, then the UF would be 1. If the study were <25% of the expected average life-span for the endangered species the UF would be 10, as could happen in human risk assessment.

3. In a practical sense it is unlikely that adequate data would exist on various endangered species. In cases of inadequate data on the species of concern, standard procedures used in human risk assessment, therefore, would be used (i.e., use of surrogate species).

G. LACK OF ACHIEVEMENT OF STEADY-STATE (SS) UF

1. Concept

It has been argued that agents requiring a long time to achieve steady-state may have their chronic toxicity underestimated in short-term aquatic (McCarty et al. 1985) and mammalian (Mosberg and Hayes 1989) toxicity studies. In order to prevent such underestimates of toxic responses, the risk assessor must assure that the compound under study has achieved approximate steady-state for the duration of the study. This will assure that the available data will have considered toxicity during both uptake (i.e., kinetic-phase toxicity) and steady-state (i.e., inherent toxicity) phases of exposure.

The concepts of kinetic-phase and inherent-phase toxicity have implications for the derivation of the ACR. The data suggest that compounds which achieve steady state (SS) quickly are more likely to have a lower ACR than those agents which take a long time to achieve SS. This suggests further that the magnitude of the average ACR is likely to be influenced by the selection of chemicals employed to construct the distribution. If the agents selected for the ACR achieve SS quickly, a lower ACR would be predicted and vice versa. It should be emphasized that factors other than achieving SS may be more important in the causation of toxicity. This discussion is not designed to minimize this possibility, but to ensure that the time to achieving SS concept is given proper consideration in the interpretation of the ACR.

While the % SS UF has a role in overall UF consideration in the process of ecological risk assessment, it is limited because of potential for interdependence with other UFs, such as the LL-UF and the ACR UF.

2. Potential Interdependence of ACR UF and % SS UF

In most circumstances it would be expected that the acute to chronic UF should have taken the % SS UF at least partially into account since some degree of contaminant uptake would have occurred. In such cases there would be potential for an error of double counting (i.e., interdependence of UFs) if both UFs are employed. Prior to determining the potential size of the % SS UF it would be necessary to estimate the % of SS achieved in the chronic study. At that point it may be possible to address the question of whether a % SS UF may be necessary and what its size should be.

3. Potential Interdependence of the LL-UF vs % SS UF

The LL UF should be partially interdependent with the % SS UF. This conclusion derives from the belief that the LL UF takes into account generally those agents that continue to accumulate, but that do not achieve SS after the end of the LL study. Thus, when a LL UF is used, the % SS UF should not be used to normalize the LL study to one of a chronic nature. The question which then emerges is: when (if at all) should the % SS UF be used?

1. It could be used in chronic studies where SS has not been achieved. However, the size of this UF would need to be addressed on a case-by-case basis.
2. The LL UF should account for the % SS UF for the time period dealing from the completion of the LL study to the completion of a chronic study; however, there would remain concern for that period of life after the completion of the normal chronic study to the average expected life span.

4. Recommendation

The only time the % SS UF should be employed is for that time period after the end of a normal chronic study (as achieved by a chronic study via the use of a LL-UF or A/C UF) when the concentration of the toxicant continues to increase in bodily tissues. However, this type of information would generally not be available and, if it existed at all, would likely be modeled predictions. Consequently, we believe that while the concept of a % SS UF is theoretically valid, it has limited utility and would normally not be employed. If it were ever to be employed, it would be on a case-by-case basis.

H. MODIFYING FACTOR (MF)

This factor will be handled in a similar fashion as seen in human risk assessment (see Table 4.1).

I. LABORATORY—FIELD EXTRAPOLATION UF

1. Rationale and Recommendation

Whether and to what extent a separate UF should be used for extrapolation from the laboratory to the field has been a much discussed (Slooff et al. 1986; Van Straalen and Denneman 1989), but essentially unresolved issue. It has been argued that laboratory studies have the potential to both under- and overestimate field-based responses (Table 4.13). Thus, a laboratory-to-field extrapolation factor, if adopted, could be either greater or smaller than one. According to Van Straalen and Denneman (1989) no convincing case has yet been made concerning the magnitude of the laboratory-to-field UF and, at least for the foreseeable future, a factor of one should be applied. If one were to be adopted, it should be approached on a case-by-case basis.

J. INTERDEPENDENCE OF UFs: AVOIDING ERRORS IN OVER-CONSERVATIVE APPLICATION OF UFs

For UFs to be properly employed it is assumed that they are independent of each other and, therefore, have a multiplicative interaction. However, several instances of interdependence of UFs exist, so that their joint use should be avoided.

1. Acute to Chronic UF and the LL UF

If an acute to chronic UF is used it is assumed that the UF will estimate the response over a normal adult life span. Thus, use of a LL UF in this instance would be an error in double counting.

Table 4.13.   Arguments Used in Establishing a Laboratory-Field Extrapolation Factor

+1. In the laboratory, organisms are tested under optimal conditions.

−2. In the field, biological availability of chemicals is lower than in laboratory tests.

+3. In the field, organisms are exposed to mixtures of many chemicals.

−4. In the field, ecological compensation and regulation mechanisms are operating.

−5. In the field, adaptation to chemical stress may occur.

+6. Adaptation often entails costs in ecological performance.

Source: Van Straalen and Denneman, 1989.
Note: A plus sign indicates a positive argument to maintain an extrapolation factor greater than one; a minus sign indicates a negative argument.

2. Acute to Chronic UF and Intraspecies (i.e., Sensitive Life-stage) UF

If the A/C UF is based on complete life-stage data then the intraspecies UF should not be employed. This concern with the interdependence of UFs would not apply to protected species. In the case of endangered species the use of the A/C UF would not preclude the use of an intraspecies UF. This is because the typical chronic stage would not address the interindividual variation as required for protection of endangered species.

3. Interspecies UF and Intraspecies UF

The interspecies UF in human risk assessment is generally recognized as providing an extrapolation from the average animal to the average human, assuming that humans may be 10-fold more sensitive. The interindividual UF assumes that most (not necessarily all) human responses to an agent fall within approximately a 10-fold range. Given this assumption, the application of a 10-fold interindividual UF should begin with the average person and extend to cover the higher risk segments of the population. Consequently, an UF of 5 would be expected to protect most humans (Figure 4.4).

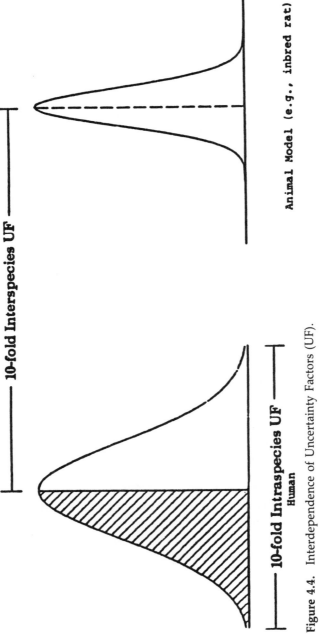

Figure 4.4. Interdependence of Uncertainty Factors (UF).

The application of a 10-fold UF for humans would be more justified if it were based on an occupational epidemiological study. This type of study does not consider the most sensitive humans and is likely to involve principally healthy workers and a self-selection component that consists of the less sensitive members of the population (Figure 4.5). Therefore, it is concluded that the current use of a 10-fold factor for interindividual variation, as typically applied to animal toxicological studies used in risk assessment, represents an important deviation from the original intention of uncertainty factor use. This intention for interindividual variation is satisfied with an UF of 5 when based on animal studies, but with a factor of 10 when based on occupational epidemiological studies. It should be emphasized that this argument would apply to the basic relationship between the interspecies and intraspecies UF. The UF values of 10 and 5 are used here because of their relationship to current EPA practice and for illustrative purposes. This concept would be applicable whether the size of the UFs were larger or smaller than 10. *Therefore, when both of these UFs are present the size of the intraspecies UF should be reduced by 50%.*

K. ALTERNATIVE TO THE USE OF UFs

In the face of mounting costs associated with remedial actions at hazardous waste sites, there is great incentive to find ways to reduce toxicological uncertainties with realistic testing so that the magnitude of traditionally combined UFs and, therefore, costs can be mitigated. With this as their goal, the Department of Defense has set forth on a program to develop fast and relatively inexpensive nonmammalian toxicity assessment techniques that can be employed not only in the laboratory, but also at field sites having contaminated water (Van der Schalie and Gardner 1989; Gardner 1990). The advantage of such a system is that data can be obtained on realistic exposures within the context of multiple dilutions with large numbers of fish per exposure level. The system is also set up to consider a broad spectrum of toxic endpoints, including immunological alterations, tumor promotion, and organ specific toxicities. The plan is unique in that NOAELs for sensitive endpoints may be obtained with the toxic mixture of concern. This type of approach has the capacity to significantly reduce some of the uncertainties associated with site assessments, especially those dealing

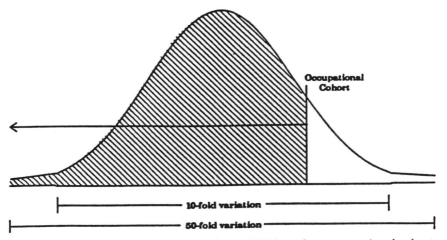

Figure 4.5. Intraspecies Uncertainty Factor (UF) based on occupational cohort.

with high to low dose concerns, identification of NOAELs, and possible additive or synergistic effects. It also can be applied so that sensitive life stages could be tested within the context of defining chronic toxicities. The methodology, however, would still need to assess the issue of interspecies uncertainty.

L. SUMMARY

In conclusion, the concept of UFs in ecological risk assessment is conceptually well established. While no widespread agreement exists concerning their magnitude, a series of recommendations for UF size (Table 4.14), along with the biological basis to support such decisions, is provided. Even though a number of similarities exist between the use of UFs for noncarcinogens in human and ecological risk assessment, major differences exist with respect to the types of UFs and their magnitude. The driving force for variation between the two schemes relates to the goal of human risk assessment to protect individuals, while the goal of ecological risk assessment is to protect populations. Other factors contributing to variability in the size of UFs between ecological and human risk assessment procedures relate to the need of ecological risk assessment to address multispecies variability and ecosystem health instead of only one species (i.e., humans). In addition, variation in

Table 4.14.   Recommended UFs in Ecological Risk Assessment

Types of UFs	Size of UFs
1. Interspecies Uncertainty	
a. Species with genus	10
b. Genera within family	30
c. Families within order	60
d. Orders within class	100
e. Classes within phylum	1000
2. Intraspecies Uncertainty	
a. Nonendangered Species (addresses developmental/reproductive endpoints)	10^a
b. Endangered Species (addresses interindividual variation including sensitive life-stages)	20^a
3. Less-than-Lifetime at Steady State	
i. $\geq 15\%$ of normal adult life span.	1
ii. $< 15\%$ of normal adult life span (but not an acute toxicity test)	10
iii. $< 25\%$ of normal adult life span for an endangered species	10
4. Acute Toxicity to Chronic NOAELS	
i. Lethality (LD to LD_{50} range)	50 at 95%
	100 at 99%
ii. If nonlethal but frankly toxicity effects occur:	$> 10 - < 50$
5. LOAEL to NOAEL	10
6. Modifying Factor	up to 10

[a]The values of the intraspecies UF will be reduced by 50% when used in conjunction with interspecies UF.

usual testing protocols has implications for how LOAEL/NOAELs may be estimated between the two approaches.

5

Deriving MATCs:
Chemical- and Species-Specific

A. INTRODUCTION

Since life cycle test data are not routinely available for a specific chemical and species, it is often necessary to employ other data to estimate the needed hazard assessment information. These data may include full or partial life-cycle data for another species or physical-chemical properties of the chemical in question. Based on such considerations, Barnthouse et al. (1990) proposed a hierarchy of toxicity data to derive a set of extrapolation models for each category. From the highest to lowest quality of potentially relevant information, the following types of data were included:

- Life cycle test for the species of interest with data for all life stages.
- A life cycle test for a species other than the species of interest.
- An incomplete chronic test for the species of interest.
- An incomplete chronic test for a species other than the species of interest.
- An LC_{50} for the species of interest.
- An LC_{50} for a species other than the species of interest.
- Physical-chemical property(ies) of the chemical.

Once the quality and quantity of the hazard assessment database have been determined it is necessary to determine how this information can be utilized in the derivation of MATCs. Within this context, the various MATC-derivation procedures previously described will be evaluated based on the role of acute-to-chronic toxicity extrapolation, inter- and intraspecies uncertainty factors, and the remaining issues in the extrapolation process.

At the outset of this process, which is designed to select the most toxicologically defensible and relevant methodology, it is important to note that verification or validation of the respective procedures for even a limited dimension is usually not available. This is typically the case for human risk assessment extrapolation procedures as well. In such cases, the most influential factor will be the biological plausibility of the assumptions within the context of assuring acceptable protection.

Another factor of critical importance is that of the endpoints of concern. All the methodologies noted above have concerned themselves within the traditional motif of ecological risk assessment by restricting themselves to endpoints affecting growth, maintenance and reproduction (e.g., hatching, larval survival, fecundity, and parental survival). They have not specifically addressed contaminant-induced pathology or endpoints such as cancer. Concerns with teratogenicity are not specifically addressed except insofar as hatch rates and larval survival are reflective of teratogenic effects. However, those parameters would not be adequate to properly address teratogenicity by themselves.

The above discussion is of relevance since endpoints such as cancer and teratogenicity that act via a genetic mechanism may occur at substantially lower dosages than those usually affecting growth, maintenance, and reproduction. In addition, the modeling that is adopted to extrapolate or predict the traditional ecological risk effects may not be the most appropriate for carcinogenic and teratogenic effects.

B. OPERATIONAL PROCEDURES: HOW TO DERIVE A SPECIES-SPECIFIC MATC

1. General Information

Barnthouse et al. (1990) propose that concentration-response functions should be fit to the data of each life stage (Box 8, Figure 5.1) and the derived equations and error associated with the model fit should be used to estimate responses to the agent in the population. They recommend the use of a logistic function;*

---

*Crump (1984) has established that the derived acceptable dosage will not be highly model-dependent since the method will not involve extrapolation far below the experimental range. In this respect, the method selection issue avoids much of the concern associated with model dependency seen with extrapolation of carcinogenicity data to low doses.

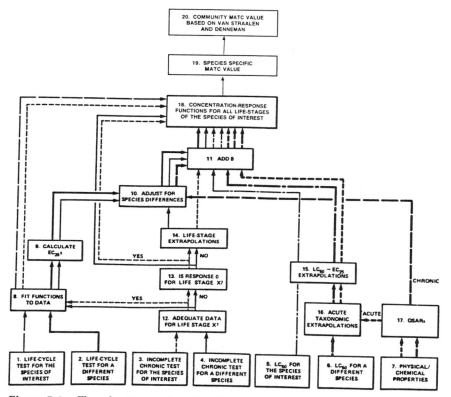

Figure 5.1. Flowchart to guide estimation of maximum acceptable contaminant concentrations from seven common types of toxicity test data. Based on scheme of Barnthouse et al. 1990.

$$P = e^{(a + Bx)}/[1 + e^{(a + Bx)}]$$

where P is the fractional response in the test population, x is the exposure concentration, a and B are fitted parameters, and B is a measure of the steepness of the function. All concentrations are expressed as $\log_{10}$ µg/L.

The mortality data are corrected for control mortality using Abbott's formula. With the exception of data box #1 (Figure 5.1), all other data boxes (Boxes 2 to 7 of Figure 5.1) require extrapolation of values a and B for one or more life stage, according to Barnthouse et al. (1987). Barnthouse et al. (1990) state that either (a) the parameters a and B from Box 8, or (b) one of the two parameters plus the EC25 are enough to uniquely define a logistic concentration-response function. Concentration-response functions are obtained by (a) estimating a chronic EC25, (b) assigning a value to B from the observed Bs for all available chronic concentration-response data sets, and (c) solving for "a" as follows:

$$a = \ln [1/3 - B(EC25)]$$

Barnthouse et al. (1990) argue that there is no compelling evidence that systematic differences exist between species or life stages in the shape of response curves that would permit anything other than only a random assignment of the B value from the observed distribution.

The Barnthouse et al. (1990) regression procedure allows the estimation of the concentration-response for any desired level of effect and is favored over approaches using hypothesis testing. Hypothesis testing compares responses at exposure concentrations with control responses to evaluate the null hypothesis. This approach has been traditionally used to derive regulatory-type benchmarks such as NOELs, LOELs, and MATCs. However, the use of the hypothesis testing approach has important disadvantages including: (1) the threshold for statistical significance does not correspond to a toxicological threshold or to any particular level of effect; and (2) poor testing procedures increase the variance in response, thereby diminishing the apparent toxicity of the agent in a hypothesis test. Hypothesis testing does have the advantage under conditions where test data are too poor and limited for curve fitting (see Stephan and Rogers 1985 for a detailed discussion). Since aquatic data usually involve a control and five treatments, the regression approach would be preferred in the majority of cases.

In the case of regression modeling, the MATC would not usually be derived. However, the MATC can be derived this way by establishing a priori the "acceptable" response level according to the biological significance of the endpoint (e.g., survival, fecundity, etc.). Within this context, one would utilize the regression equation to estimate the concentration for the desired response (i.e., passing from Box 18 to 19).

2. Interspecies (Taxonomic) Extrapolation (Boxes 2, 4, and 6 of Figure 5.1)

a. Option 1:

This estimation would ideally be performed by a direct extrapolation of acute and chronic responses between the test species and the species of interest, as is done with LC_{50} data. However, since the data on chronic concentration-responses are available for only a small number of species, Barnthouse et al. (1990) recommend the use of the acute taxonomic extrapolation because the relative sensitivity of fish taxa are generally assumed to be the same in acute and chronic tests. This assumption is based on the observations that the acute-to-chronic ratios are similar regardless of the inherent susceptibility of the species (Kenaga, 1982). Once the chronic sensitivities are estimated, then addition of the slope value (B) will define the function (Box 11 of Figure 5.1, Table 5.1). While the risk assessor would have the option of selecting the best estimate or any desired prediction interval estimate (e.g., 90, 95, 99), we recommend using the 95% level.

b. Option 2:

The use of generic UFs is considered a legitimate procedure for use in ecological risk assessment. This procedure is based on the binary comparisons of Suter et al. (1987) and Barnthouse et al. (1987, 1990), who define the regression equation. In a general sense, this approach is more consistent with the traditional notion of UFs and has more conservatism built into its estimate than the tailored UF of Option 1.

This approach makes use of taxonomic UFs, the size of which is a function of phylogenetic relatedness as given in Table 4.10.

3. Intraspecies (Life Stage) Extrapolation (Boxes 3 or 4)

a. Option 1:

i. Response Function: Barnthouse et al. (1990) indicate that incomplete chronic data for the species of interest will likely occur. Usable incomplete chronic data exist only when a partial chronic test (e.g., early life stage test) is available and when the data satisfy criteria for regression analysis (Box 12 of Figure 5.1, Table 5.1). Barnthouse et al. (1990) argue that the data are not satisfactory: (1) if there is greater than 30% mortality, (2) if there is less than 25% reduction in performance relative to controls at the maximum concentration studied, or (3) if the confidence interval on B includes zero (i.e., no significant increase in response within the range of tested concentrations observed).

Table 5.1.   Extrapolation Equations for Estimating Chronic Effects (Barnthouse et al. 1990)

No.	X Variable	Y Variable	Slope	Intercept	n	Prediction Interval[a]
Life-stage extrapolations						
1.	MORT2[b] EC25	MORT1[c] EC25	1.2	−0.12	16	0.89
2.	MORT2 EC25	HATCH[d] EC25	1.0	0.18	28	1.1
3.	HATCH EC25	MORT1 EC25	0.80	−0.05	7	1.1
4.	HATCH EC25	EGGS[e] EC25	0.78	0.11	6	1.1
5.	EGGS EC25	MORT1 EC25	1.0	0.48	27	1.2
6.	EGGS EC25	MORT2 EC25	0.88	0.48	25	0.84
Acute-chronic extrapolations						
7.	LC_{50}	HATCH EC25	1.1	−1.2	31	1.7
8.	LC_{50}	MORT1 EC25	0.87	−0.87	28	1.5
9.	LC_{50}	MORT2 EC25	1.0	−0.89	89	1.5
10.	LC_{50}	EGGS EC25	1.1	−1.89	42	1.8
Taxonomic extrapolations: Species within genera						
11.	Salmo clarkii	S. gairdneri	0.98	0.04	18	0.96
12.	Salmo clarkii	S. salar	1.0	−0.25	6	0.78
13.	Salmo clarkii	S. trutta	1.0	−0.20	8	0.74
14.	Salmo gairdneri	S. salar	1.2	−0.51	10	0.87
15.	Salmo gairdneri	S. trutta	1.1	−0.21	15	0.56
16.	Salmo salar	S. trutta	1.0	0.09	7	0.70
17.	Ictalurus melas	I. punctatus	1.1	−0.11	12	0.66
18.	Lepomis cyanellus	L. macrochirus	1.1	−0.62	14	0.80
19.	F. heteroclitus	Fundulus majalis	1.1	−0.32	12	0.75
Taxonomic extrapolations: Genera within families						
20.	Oncorynchus	Salmo	1.0	−0.13	56	0.65
21.	Oncorynchus	Salvelinus	1.1	−0.47	13	0.57
22.	Salmo	Salvelinus	1.1	−0.33	56	0.73
23.	Carassius	Cyprinus	1.0	−0.47	8	0.58
24.	Carassius	Pimephales	1.0	−0.27	19	0.82
25.	Cyprinus	Pimephales	0.93	0.24	10	0.82
26.	Lepomis	Micropterus	1.0	−0.20	30	0.92
27.	Lepomis	Pomoxis	0.82	−0.01	8	0.94
28.	Cyprinodon	Fundulus	0.96	0.21	12	0.75
Taxonomic extrapolation: Families within orders						
29.	Centrarchidae	Percidae	0.95	−0.02	47	1.01
30.	Centrarchidae	Cichlidae	0.93	0.40	6	0.56
31.	Percidae	Cichlidae	1.4	0.15	5	1.12
32.	Salmonidae	Esocidae	1.4	−0.49	11	0.94
33.	Atherinidae	Cyprinodontidae	0.90	0.50	32	0.83
34.	Mugilidae	Labridae	0.82	0.70	12	1.74
35.	Cyprinodontidae	Poecillidae	0.75	0.19	12	0.53

continued

Table 5.1. *Continued*

No.	X Variable	Y Variable	Slope	Intercept	n	Prediction Interval[a]
Taxonomic extrapolation: Orders within classes						
36.	Salmoniformes	Cypriniformes	0.87	0.90	225	1.31
37.	Salmoniformes	Siluriformes	0.85	0.87	203	1.59
38.	Salmoniformes	Perciformes	0.94	0.33	443	1.09
39.	Cypriniformes	Siluriformes	0.93	0.23	111	1.04
40.	Cypriniformes	Perciformes	0.99	−0.39	219	1.51
41.	Siluriformes	Perciformes	1.1	−0.74	190	1.8
42.	Anguiliformes	Tetraodontiformes	0.89	1.09	12	1.1
43.	Anguiliformes	Perciformes	0.96	0.21	34	1.4
44.	Anguiliformes	Gasterosteiformes	1.0	0.52	8	1.2
45.	Anguiliformes	Atheriniformes	1.0	0.06	46	0.94
46.	Atheriniformes	Cypriniformes	0.82	1.93	7	2.7
47.	Atheriniformes	Tetraodontiformes	0.88	1.00	46	1.1
48.	Atheriniformes	Perciformes	0.92	0.10	148	1.4
49.	Atheriniformes	Gasterosteiformes	0.94	0.49	36	1.3
50.	Gasterosteiformes	Tetraodontiformes	1.12	0.31	8	1.3
51.	Gasterosteiformes	Perciformes	1.15	−0.67	33	1.5
52.	Perciformes	Tetraodontiformes	0.91	0.93	34	1.4
Taxonomic extrapolation: All fish from standard species, within media						
53.	*P. promelas*	Osteichthes	1.01	−0.30	354	1.3
54.	*L. macrochirus*	Osteichthes	0.96	0.17	500	1.4
55.	*S. gairdneri*	Osteichthes	0.99	0.29	480	1.2
56.	*C. variegatus*	Osteichthes	0.97	0.03	51	1.5

Units are log $\mu g/L$.
[a]The 95% prediction interval at the mean is log $Y \pm$ the number in this column.
[b]Mortality of larval fish.
[c]Mortality of parental fish.
[d]Fraction of eggs failing to produce normal larvae.
[e]Number of eggs produced per female fish surviving to the beginning of spawning.

ii. Tailored Life Stage Extrapolation: In the case where there is no response within the range of concentrations tested (Box 13), Barnthouse et al. (1990) assigned a zero effect to that life stage. If the data for a life stage are inadequate because the control mortality is greater than 30%, or if it is missing, then they recommended life stage extrapolations (Box 14). This step is performed by the use of regressions to estimate EC25 values for the missing (or inadequate) data based on data for life stages with adequate data. How this is accomplished is described by Barnthouse et al. in the following example. Assume data exist only on larval mortality. The

EC25 for larval mortality would then be used to estimate the EC25 for parent mortality and fecundity based on equations (Table 5.1) for various life stage extrapolations. Slope (B) values would be assigned to the EC25 values (Box 11), thereby providing a full set of concentration-response functions.

In the case of Box 4, a double extrapolation procedure is required because it represents an example of incomplete chronic data for a different species. This situation is dealt with as a combination of the inter- and intraspecies extrapolation procedures. According to Barnthouse et al. (1990), functions would be fit to those life stages with adequate concentration-response data (Boxes 12 and 8), EC25s would then be determined (Box 9) with an adjustment made for interspecies variation (Box 10), and the slope function (B) added to complete the functions (Box 11). In the case of life stages with no response within the concentrations evaluated, a response of zero is given (Box 13). For life stages where data do not satisfy a priori criteria, but where the response is not zero, the response is estimated via life stage extrapolation (Box 14), and the resulting EC25s are adjusted for species variation (Box 10) and given a B value.

b. Option 2:

In a comparable manner to Option 2 in the interspecies extrapolation section, an UF factor approach may be considered. Although it has not been used or recommended in the published literature as yet, the technique of Slooff et al. (1986) for derivation of a 95% UF using regression analysis would be a legitimate approach. This approach has the same advantages as those noted above and would also be derived in an identical manner. The magnitude of such a UF would, of course, be dependent on the database. In addition, a 99% or 90% upper or lower prediction interval (PI) estimate could be used, thereby providing more flexibility to the risk assessor.

An alternative to the tailored UF approach is the use of a generic intraspecies UF. Review of the data provided by Barnthouse et al. (1987) indicates that the majority of variations in response will be within one of magnitude (see Table 4.12). This is very comparable to a large mammalian data set assembled by Calabrese (1986) which showed that age-related sensitivities were log-normally distributed, with mean responses being approximately 10-fold while median values were slightly above 3-fold. Such information suggests that a generic factor of 10 for a life-stage UF would be protective of most, but not all, comparisons.

This type of double extrapolation procedure of Barnthouse et al. (1990) can be readily achieved by combining the directives under Option 2 for both inter- and intraspecies extrapolation. In the case of using a 95% UF, inter- and intraspecies values would then be independently estimated and then multiplied to provide the total size of the uncertainty factors. This combined value would then be used to estimate the MATC for a specific species in Box 19.

4. Acute-to-Chronic Extrapolation

a. Option 1:

In the case where only acute toxicity values (i.e., 96 hr LC_{50}) exist for the species of interest, Barnthouse et al. (1990) recommend the use of LC_{50}-to-EC25 extrapolations (Box 15 of Figure 5.1, Table 5.1). Such acute-to-chronic (or high-to-low dose) extrapolation is accomplished by regressing EC25 values for each life stage against LC_{50} values for the same species, chemical, and test conditions. Values are then assigned to the EC25 to complete the function. When acute toxicity data exist for a different species (Box 6), Barnthouse et al. (1980) recommend employing acute taxonomic extrapolations to estimate the LC_{50} for the species of interest (Box 16). As in previous cases, these extrapolations between fish taxa estimate LC_{50}s for other fish species from the LC_{50} of standard test species. The taxonomic extrapolations are conducted between taxa having the next higher taxonomic level in common. The LC_{50}s, which are now adjusted for the relevant species, are employed in the LC_{50}-to-EC_{25} extrapolation procedure (Box 15) and the B value is added. While the risk assessor would have the option of selecting the best estimate of any desired prediction interval estimate (e.g., 90, 95, 99), we recommend using the 95% level.

b. Option 2:

As in the previous cases of extrapolation, use of a 95% UF factor may be employed so that a generic UF can be estimated from regression equations used in Option 1 of this acute-to-chronic extrapolation section (Barnthouse et al. 1990). In this case the size of the UF would be specific to the database from which it is derived and could be adjusted to any upper or lower bound value (e.g., 90%, 95%, 99%). Table 4.3 indicates that the generic acute-to-chronic UF would be 50-fold for the 95% level and 100-fold for the 99% prediction interval. This factor would be multiplicative and combined with previous UFs and employed in Box 19.

5. Extrapolation for Structure Activity Relationships

While Barnthouse et al. (1990) proposed a scheme to estimate concentration response functions for all life stages of the species of interest, it is recommended that these procedures not be adopted. Instead, it is recommended that actual testing data be required. While reasonable schemes could be and have been derived in some chemical classes with respect to making biological/toxicological predictions based on QSAR, it is believed that extensive uncertainty already exists and that this further extension should not be followed. This is especially the case since preliminary biologically-based data of acute or life stage data are readily obtainable.

C. SKELETONIZED PROCEDURES: HOW TO DERIVE A CHEMICAL-SPECIFIC ECOSYSTEM EXPOSURE CRITERION

1. Situations

1. Life-cycle test for the species of interest:
   A. Derive MATC at response appropriate (e.g., EC10, EC25) for the biological relevance of the endpoint
2. Life-cycle test for a different species
   A. Derive preliminary MATC at response (e.g., EC10, EC25) appropriate for the biological relevance of the endpoint
   B. Apply interspecies (taxonomic) UF Options for selection of UF
      i. Use taxonomic extrapolation scheme of Barnthouse et al. (1990) and select the upper bound (95%) of the prediction interval if the database is relevant [i.e., if species of interest is in the Barnthouse et al. (1990) database]; if it is not relevant, then use the generic guidance of Slooff et al. (1986) and Barnthouse et al. (1990) and apply an UF; the size of the UF would depend on the degree of phylogenetic relatedness (Table 4.14).
      ii. Apply the UF to the preliminary MATC value

3. Incomplete chronic test for the species of interest
   A. Use life stage extrapolation procedures of Barnthouse et al. (1990) to provide best estimate of MATC if the data are available and relevant.
   B. If life stage extrapolation procedures from Barnthouse et al. (1990) are not available or appropriate, use a regression analysis from the experimental data to estimate a preliminary MATC. Then apply a UF of 10 (Table 4.14) to derive the final MATC for this species of interest.
4. Incomplete chronic test for a different species
   A. Approach A
      i. Use life stage extrapolation procedures of Barnthouse et al. (1990) to provide best estimate for life cycle response.
      ii. Use taxonomic extrapolation procedure of Barnthouse et al. (1990) to provide best estimate for species of interest; this will provide the MATC value.
   B. Approach B
      i. If life stage extrapolation procedures of Barnthouse et al. (1990) are not available or relevant, then a life stage UF of 10 (Table 4.14) can be applied to a subsequently derived preliminary MATC.
      ii. If taxonomic extrapolation procedures of Barnthouse et al. (1990) are not available or relevant, then a taxonomic UF depending on phylogenetic relatedness (Table 4.14) can be applied to the preliminary MATC value.
      iii. Both UFs will be multiplied, yielding a final UF which can be divided into the preliminary MATC.
      iv. The preliminary MATC can be derived by the use of regression analysis of the experimental data at a response appropriate for the biological relevance of the endpoint.
   C. Approach C
      i. If only taxonomic data are available and relevant from Barnthouse et al. (1990), then use the above procedure to derive a preliminary MATC.
      ii. Next derive two UFs. The life stage UF is a generic 10-fold value. The taxonomic UF is derived from the 95% value of the prediction interval of the taxonomic regression extrapolation.
      iii. The UFs are then multiplied and the resultant product divided into the preliminary MATC to derive a final MATC.
4. LC_{50} for species of interest
   A. Approach A
      i. Use acute-to-chronic extrapolation procedures of Barnthouse et al. (1990) to derive a preliminary MATC.

  ii. Derive an acute-to-chronic UF based on the 95% prediction interval.
  iii. Divide this UF into the preliminary MATC to obtain a final MATC.
  B. Approach B
  i. If data from Barnthouse et al. (1990) or equivalent data are not relevant or available, then apply an UF of 50-fold (95% prediction interval) or 100-fold (99% prediction interval) to the acute value.
5. LC_{50} for a different species
  A. If data on acute taxonomic extrapolation from Barnthouse et al. (1990) or equivalent data are available and relevant, then perform a regression analysis that will produce a tailored UF using the 95% prediction interval. If these data are not available or relevant, then a generic taxonomic UF based on phylogenetic relatedness (Table 4.14) can be subsequently applied to the preliminary MATC.
  B. If an acute-to-chronic extrapolation procedure is not available, then the generic UFs for acute-to-chronic (Table 4.14) and taxonomic variation based on phylogenetic relatedness (Table 4.14) can be multiplied to yield a final UF. This final UF would then be divided into the LC_{50} of the different species.

D. STRENGTHS AND LIMITATIONS OF THE RECOMMENDED METHODS (See Appendix 2 for specific examples of how these methodologies are applied.)

1. Strengths

1. A wide range of toxicity data can be used.
2. Biological plausibility is generally strong.
3. Regression analysis is utilized rather than hypothesis testing.
4. The methods are not technically complex.
5. UF factors are obtained from experimentally derived data.
6. All major sources of uncertainty are considered: interspecies, intraspecies, and acute to chronic.
7. MATC values are developed as a function of biological relevance of endpoint.
8. Alternative approaches are generally consistent with recommended processes.
9. Studies with relevant data are rewarded.
10. Well-conducted studies are rewarded.

2. Limitations

1. Adequate validation is lacking.
2. Use of the regression model assumes sufficient data.

6

The Toxicity Reference Value (TRV)

A. WHAT TO DO IF THE MATissueC APPROACH CANNOT BE EMPLOYED?

There are circumstances when the MATissueC approach cannot be readily or directly adopted, given its inherent predictive limitations and need for external corroboration. For example, tissue samples from individuals of endangered species and migratory birds are not likely to be available for use in MATissueC derivation. Agents that do not bioaccumulate do not lend themselves to the strengths of the MATissueC/food web modeling approach. Limited resources may also restrict the capacity to obtain data on species of concern and appropriate environmental media, or to perform complex food web analyses. In such data-limited situations, it may be necessary to develop surrogate estimates of exposure to replace (or to complement) complex food web modeling estimates, and to determine accepted exposure rates (to replace tissue level concentrations) on species that are not those of concern. This type of situation presents difficult challenges to the ecological risk assessor because the lack of both quality and quantity of data creates a situation which is bounded by considerable uncertainty.

The risk assessor will need to derive the equivalent of a mammalian reference dose (RfD) for terrestrial species. Acceptable intake rates for terrestrial animals can be guided by the premise that ecological risk assessments are designed to protect the population and not individual response, except in the case of specially protected species as mandated by legislative acts.

The type of endpoints that ecological risk assessments need to address for nonendangered species include growth, maintenance, reproduction, and critical generic developmental processes. Chronic diseases such as cancer which usually affect older individuals are not typically selected as endpoints upon which to base a chronic exposure rate. However, cancer or other chronic diseases may be appropriate endpoints to consider for reference exposure derivation for endangered species.

The derivation of an acceptable chronic exposure rate for terrestrial species in ecological risk assessment, as well as in human risk assessment, must deal with various types of uncertainties. These include interspecies variation, intraspecies (insofar as developmental and reproductive processes are considered), FEL (frank effect level), LOAEL-to-NOAEL extrapolation, and less-than-lifetime (although with much greater restrictions than used in human risk assessment). Such restrictions need to be consistent with the goals of population versus individual protection.

B. CURRENT ATTEMPTS TO DERIVE TRVs

Limited regulatory attention has been given to deriving health-based terrestrial (i.e., nonhuman) receptor toxicity reference values. The largest initiative in this regard is that being developed with respect to the RMA by contractors to DOD (Fordham and Reagan, 1991). These efforts have resulted in an approach for the endangerment assessment (EA) at the RMA that is claimed to be based on the EPA approach for deriving human RfDs. The acronym RfD, however, has been replaced by TRV (i.e., Toxicity Reference Value). Even though there is a change in name, the methodology deriving TRVs employs the same concepts used by EPA in the RfD development (HLA, 1991). Despite the claim that exposure estimates (mg/kg/day) derived by the TRV method are "without an appreciable risk of deleterious effects to that population," no definition of appreciable risk is given. Endangered species are provided with an additional UF in the form of a modifying factor (MF). *The HLA report (1991) indicates that any differences that arise between the TRV versus the RfD processes result from the intent of the RfD to usually protect the most sensitive individuals rather than to protect the population as a whole, as is the intent of ecological risk assessment.* Figures 6.1 and 6.2 provide the decision-tree framework to derive TRVs.

The HLA (1991) report argued that exposure rates protective of terrestrial populations need not be as low as that required to help ensure protection of individual humans. As a result, the magnitude

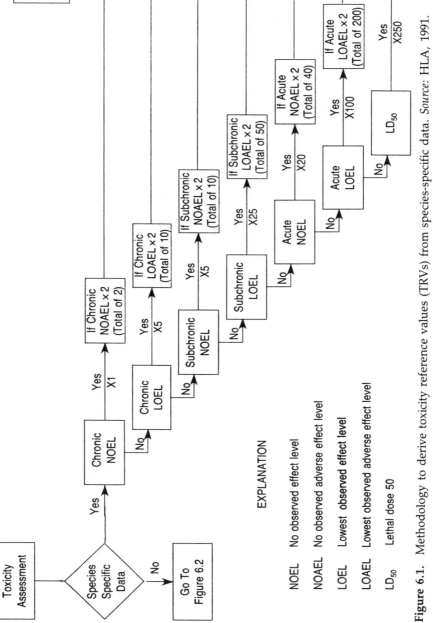

Figure 6.1. Methodology to derive toxicity reference values (TRVs) from species-specific data. *Source:* HLA, 1991.

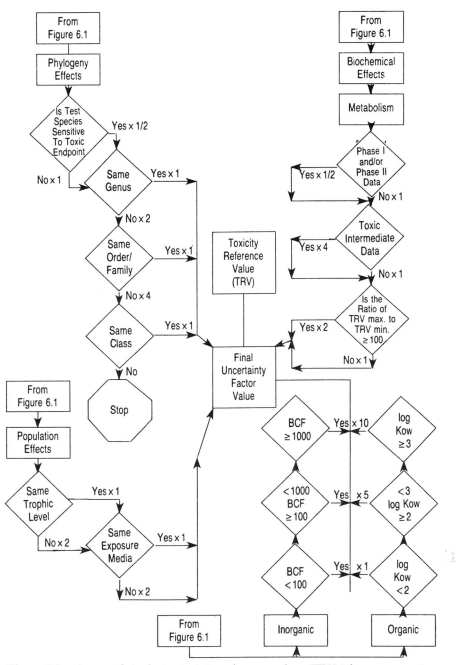

Figure 6.2. Approach to derive toxicity reference values (TRVs) from nonspecies-specific data. *Source:* HLA, 1991.

of the UF's in human risk assessment, which often employs values up to 10-fold for each of 1 to 5 uncertainty areas (e.g., species variation, interindividual variation, less-than-lifetime exposures, LOAEL to NOAEL), was deemed as unnecessarily conservative. No data were offered in the HLA (1991) report to support this (i.e., "unnecessarily conservative") conclusion.

The next step, therefore, was to develop modifications of EPA RfD UFs that would still be reasonable, but not unnecessarily conservative (HLA, 1991). Each decision point in the TRV process was identified and assigned an UF. Since there were so many decision points (i.e., 10), a maximum of 10 for each decision point was deemed too high. However, a lower value of 2 applied to each appropriately equally weighted step would result in an "adequately represented uncertainty associated with the reported effects."* Again, there were no data presented to support the decision.

The decision-tree-type processes to derive terrestrial (TRV) criteria put forth by the HLA (1991) report are given in Figures 6.1 and 6.2. The initial key decision point for the TRV process is whether or not there are adequate toxicity data on the target species. The following section represents a critique of these two proposed methodologies (i.e., Methodology #1 TRVs for species-specific data, Methodology #2 TRVs for nonspecies-specific data). The proposed methodologies of HLA (1991) are critiqued here principally as examples of what has been proposed at a major DOD-superfund site (i.e., the Rocky Mountain Arsenal). Examination of the methodology offers a useful opportunity to explore a variety of options that could be considered in ecological risk assessment, and provides the toxicological basis for accepting or rejecting specific courses of action.

C. CRITIQUE OF TRV METHODOLOGIES

1. TRVs for Species-Specific Data (Methodology #1)

A review of methodology #1 (Figure 6.1) has resulted in an identification of the following critical issues:

---

*The only exception was an UF of 10 for chemicals with BCF >1000 or if the log Kow was ≥3. The rationale for the deviation from the use of 2 for UF size was based on the use of log descriptions of both the BCF and the Kow. If the BCF values were ≥100, ≤1000 a UF of 5 was used. A BCF of <100 had UF of 1. In a similar manner, if the log Kow was ≥2 <3, UF of 5 was employed; if the log Kow was <2, UF of 1 was used.

a. Extrapolation of NOAELs-to-NOELs and LOAELs-to-LOELs

The scheme (Figure 6.1) indicates that an UF of 2 is applied whenever a NOAEL or LOAEL is obtained. The use of an UF of 2 is at variance with how the Environmental Protection Agency and the National Academy of Sciences employ UFs for human risk assessment. For example, in such cases no UF is usually applied to a NOAEL. If a change occurs which is not an adverse effect, then little justification at present exists to use any additional UF for such purposes.

b. Magnitude of UF for Subchronic-to-Chronic Study of 5 and the LOEL-to-NOEL of 5

Figure 6.1 reveals that UFs are proposed for studies which have subchronic but not chronic data as well as for those with a LOEL but not a NOEL. These approaches are qualitatively consistent with what is currently employed in the RfD process. The usual scheme in mammalian UF application, however, is to use factors of up to 10-fold in both cases. A multiplicative expression of these two different sizes of UF would yield 25-fold vs. 100-fold UFs. The HLA (1991) report provided no clear documentation for the selection of UF size nor the basis of its specific deviation from values recommended in the RfD derivation process.

c. Different UFs for Apparent Similar Responses

The chronic LOEL received an UF of 5, while a chronic NOAEL received an UF of 2. It is not clear if there is any toxicological difference between the two terms. If a difference exists, then it should be made clear and justified. If no difference exists, then there is no justification for a different UF value for these two conditions. This situation is also the case for the subchronic LOEL and subchronic NOAEL where the UFs are 25 and 10, respectively, and acute LOEL and acute NOAEL, where the UFs are 100 and 40, respectively.

d. The Size of an Acute NOEL UF of 20

The size of UF factors for acute NOEL-to-subchronic NOEL is 4, and the size of the UF when extrapolating from a subchronic NOEL-to-chronic NOEL is 5. The cumulative total is 20-fold. The magnitude of UF for an acute LOEL (LOAEL)-to-chronic LOEL (LOAEL) is also 20-fold.

There is no documentation to support the derivation of these values. At this point, it becomes critical to know how these values are statistically determined and what biological endpoints are considered.

e. Extrapolation of LD_{50} to Chronic NOEL

The size (i.e., 250-fold) of this value is again not documented in HLA (1991). However, this is likely to protect against toxicity from approximately 99% of agents tested in the same species in a standard study (see Chapter 5).

f. Inconsistencies with Aquatic Toxicity Assessment

The magnitude of UFs in comparable extrapolations is different between the aquatic and terrestrial schemes developed for DOD by HLA (1991). For example, an UF for acute-to-chronic value is 100-fold for aquatic animals, yet 250-fold for terrestrial animals (HLA, 1991). In addition, the extrapolation difference between the acute LOEL-to-chronic LOEC is 10-fold for aquatic animals, yet 20-fold for terrestrial animals. Although the basis for such differences was not addressed, it is important to clarify.

2. TRVs for Nonspecies-Specific Data (Methodology #2)

The second methodology considers four specific decision points (Figure 6.2). These include: (1) phylogenetic relatedness of the species of concern to the species on which data exist; (2) population effects that may result from either trophic level differences and/or differential access to various environmental media; (3) metabolic factors such as information on biochemical, physiological, and pharmacokinetic effects; and (4) the probability that the agent will accumulate in the environment and/or food web and thereby reduce the success of a population. Each of these factors will be considered below.

a. Phylogenetic Relatedness

That phylogenetic relationships could assist in both the selection of the most appropriate surrogate model for interspecies extrapolation, as well as in deriving the size of an interspecies UF, is a plausible position. However, this concept is preliminary and is based on current understandings of the causes of interspecies similarities and differences in response to toxic substances (Calabrese, 1991). To date, no approach has been made by regulatory agencies to use evolutionary relationships to select models nor to determine the size of the interspecies UF. Models such as the mouse, rat, rabbit, gerbil, hamster, cat, dog, or sheep are generally considered equally ''valid'' for use in human risk assessment, regardless of phylogenetic relatedness to humans. Thus, the phylogenetic dimension used by the HLA (1991) report represents a major deviation from the RfD process. No data were presented to secure a basis for the approach, except an obscure reference by Best (1983) on the concept of transphylogenic similarities and differences with particular reference to the use of planarians for toxicological evaluations.

The selection of the size of the UF based on phylogenetic factors in the HLA (1991) report was such that if both species were of the same genus (or species), a UF of 1 was used; if the split occurred at the family or order level, a UF of 2 was used; if the animals were of the same class, a factor of 4 was applied. Phylogenetic variation larger than class was considered too great (i.e., uncertain), and no further evaluation was used (See Appendix 1 for a summary of vertebrate phylogenetic classification.) No justification was offered for this proposed quantitative scheme for size of the interspecies UF based on phylogenetic factors.

The phylogenetic approach used by HLA (1991) is an attempt to find an alternative for the interspecies UF that would be biologically rational, have some toxicological basis, and would yield a lower-sized UF than the RfD process. The principal problem with this approach is that it has not been adequately developed for risk assessment purposes. Consequently, it becomes a subjective judgment that conforms more to the original requirement for a lower-sized UF than to a process consistent with data. It should be noted that other investigators, especially Suter et al. (1983a), Slooff et al. (1986) and Barnthouse et al. (1990) have quantitatively considered the issue of toxicity susceptibility and phylogenetic relatedness. Chapter 5 provides an approach with supporting data on how the size of the interspecies UF may vary according to phylogenetic relatedness.

Another interspecies extrapolation issue was considered in the HLA (1991) report, dealt with the potential unique sensitivity/

refractoriness of the test species. If a species (i.e., species A) were selected for extrapolation to the species of concern and the species (i.e., species A) were known to be sensitive to the desired or expected effect, a UF of 0.5 was used. Conversely, if species A were considered insensitive to the expected effect, a UF of 2 was applied. The question of comparative susceptibility is an extremely complicated issue and if a decision-making scheme is to be used, then objectively derived criteria are required. Unless this is done, the approach becomes excessively subjective and unacceptable. Since no scheme with objective criteria based on supportive data was presented, this approach cannot be accepted.

While the HLA report (1991) did not provide data to support their position, considerable data exist in the open literature which support the opposite position. Blanck (1984) addressed the question of differential species sensitivity experimentally. More than 100 inorganic and organic contaminants (e.g., metals, detergents, algi-, fungi-, herbi-, insecticides, chlorinated hydrocarbons, etc.) were tested among more than 30 aquatic species. Based on the results of this study, Blanck concluded that ''no species was generally sensitive or insensitive to all chemicals.'' Such evidence that species differ in their selective sensitivity to chemicals is widespread (Patrick et al. 1968; Linden et al. 1979; Adema and Vink 1981; Kenaga 1981; Bringmann and Kuhn 1979, 1980a, 1980b, 1981; Slooff et al. 1983; Blanck et al. 1984a).

b. Population Effects

i. Trophic Level Differences: This aspect of the population effects section is concerned with a comparison of trophic levels between the experimental animal and the target species (HLA, 1991). If both animals are located within the same trophic level, an UF of 1 was employed. If the animals are at different trophic levels, then an UF of 2 was used.

This UF is different from that encountered in the RfD process since it represents an exposure factor and is based on the premise that agents biomagnify. While it is essential for exposure factors to be considered and quantified, this approach represents an unnecessary UF if, in fact, actual exposure can be reasonably estimated.

Alternative approaches have long been used to address uncertainty in exposure estimates in the RfD process that may be used in the TRV scheme. For example, estimates of water and soil consumption are based on upper 90% to 95% distributional values.

No criteria were provided in the HLA report for decision-making on trophic level factors. For example, it would not be uncommon for a species of interest to be present in multiple food webs and occupy different trophic levels in the respective food webs. In addition, the use of a trophic level UF would seem irrelevant if the compound in question did not have the potential for appreciable biomagnification.

ii. Exposure Medium or Portion of Time in a Particular Medium Contributing to Uptake: The HLA (1991) approach addressed three issues, including (1) multimedia exposure, (2) laboratory-to-field uncertainty, and (3) bioavailability. Each represents a separate technical issue and it is improper to lump them together as in the case of the HLA (1991) report. Likewise, the HLA report (1991) proposed the use of an UF of 2 if uncertainty existed in any of these three areas.

1. *Multimedia Exposure:* These are generally viewed as exposure factors, and the amount of pollutant from each source is summed in an additive fashion. This has been dealt with in varying ways in the human risk assessment process. For example, exposure to lead is quantified by media source. Drinking water has been usually considered (in the absence of specific knowledge) to be 20% of the total exposure.

2. *Laboratory-to-Field Extrapolation:* The use of a laboratory-to-field UF has been widely discussed (Okkerman et al., 1991). No convincing case yet exists to support the use of a generic UF for this uncertainty area, since factors enhancing or mitigating risk components in this relationship may be highly variable. The issue of a laboratory-to-field UF at present would need to be developed on a case-by-case basis.

3. *Bioavailability:* The issue of bioavailability via the gastrointestinal tract has been used within the RfD procedure and is typically included as a MF [e.g., various cyanide compounds in the Integrated Risk Information System (IRIS)]. The magnitude of MF employed should ideally reflect actual differences in bioavailability. If this is not possible, it is generally assumed that lipophilic agents are absorbed with an efficiency of 100%. However, this assumption must be viewed as a default option. For example, Kester et al. (1992) have recently reported that the bioavailability of aldrin and dieldrin can decrease with time of association

("aging"). These authors reported that when these pesticides were freshly applied, their relative absorption factor (RAF) (i.e., ratio of % pesticide absorbed from soil to % absorbed from diet) reduced from 94% (fresh soil) to 81% (aged soil) for dieldrin, and from 85% (fresh soil) to 40% (aged soil) for aldrin.

The combining, therefore, of these three areas (i.e., multimedia, laboratory to field, and bioavailability) to assess uncertainty under the same heading and with a collective default value of 2 is inappropriate, due to their basic inherent technical differences.

c. Biochemical Effects of the Chemical

i. Partitioning of the COC into Body Compartments (Persistence UF): This partitioning factor is another case of an exposure component. If the BCF is >1,000 for inorganics or if the log Kow for organics is >3, a UF of 10 was applied. Lower UFs were employed for small BCFs and log Kow values, as noted in Figure 6.2. The HLA report (1991) provided no rationale for selection of these UF values. Agents such as dieldrin, DDT, chlordane, and endrin have log Kows that may far exceed the Kow value of 3, but no differentiation was made for such compounds. Of further concern is that this approach is based on the premise that agents with high persistence present special environmental health concerns, and require a unique UF as well. The HLA report (1991) has opted to select the Kow for organic compounds or the BCFs for inorganic compounds as predictive indicators of the capacity to bioaccumulate in food pathways. If the exposure assessment makes use of food web modeling data, then the use of a persistence UF is redundant. If the risk assessor has no (or very poor) quantitative concept of what exposure is likely to be, the addition of an UF to a toxicity-based value is not the preferred way to deal with this concern. Again, this information should be considered in any exposure modeling, and the proposed scheme discussed above represents a redundancy in exposure assessment.

ii. Metabolism: Phase 1 and/or Phase 2 Data: The HLA (1991) report argues that metabolic data should be quantitatively used in the TRV derivation process. They stated that if data exist concerning either Phase 1 or 2 metabolism, then an UF of 0.5 should be used because the data represent a reduction in the uncertainty associated with the available toxicity data. They further state that if

Phase 1 metabolism enhances the likelihood of excretion (i.e., reduces the half-life), then the UF should be reduced. If such metabolism forms a toxic intermediate, an UF of 4 should be used to account for the uncertainty associated with the potential toxic effect.

This concept as currently proposed makes little toxicological/risk assessment sense since dose-response relationships (mg/kg/day) inherently take such metabolic factors into account. In addition, these factors may or may not be related to interspecies differences in susceptibility. The fact that Phase 1 or 2 metabolism is present, that the half-life of the agent is altered, and that the toxicity may be mediated through an intermediate are taken into account in the hazard assessment. This proposed scheme is another example of unnecessary double counting, as well as possible undercounting (UF of 0.5). This scheme is not well founded in toxicological terms, does not address interspecies variation, and is therefore not recommended.

iii. Range of Toxicity: The HLA (1991) report argued that if the range of values in the literature for similar toxicological endpoints exceeds 100, then an UF of 2 is applied. However, if the range of values is more tightly distributed (i.e., less interspecies uncertainty), an UF of 1 is used. This represents another unique approach which has no analog in the RfD process. Furthermore, it does not address issues involving numbers of species tested, comparability of studies, the most relevant species for extrapolation purposes, etc. As a result, it is lacking in criteria and justification.

D. SUMMARY

The proposed methodology to derive TRVs by the HLA report (1991) cannot be used at present because of a lack of documentation to support recommended procedures. It also suffers from a serious problem of redundant consideration of uncertainty areas, as well as a confused intertwining of hazard and exposure assessment. Despite this severe limitation in the HLA (1991) approach it was included here because it is proposed for use at the Rocky Mountain Arsenal (RMA) by the Department of Defense (DOD) and represents a major initiative in ecological risk assessment. However, its most important value is that of exploring areas of consideration in a newly emerging area and *not* its acceptance. How the TRV—UF process is applied to COCs at the RMA by DOD using the procedures of the HLA report (1991) is given in some representative examples in Table 6.1.

E. COMPARISON OF THE MATissueC VS TRV APPROACHES

The Hazard Assessment component of the TRV process establishes an acceptable exposure rate (mg/kg/day). In human risk assessment terms, exposure is generally dealt with by estimating how much contaminant is ingested from water consumption, air inhalation, food consumption, soil ingestion, etc. The summated exposure rate (mg/kg/day) is then compared to the allowable exposure rate (RfD for human risk assessment). Exposure quantification with respect to terrestrial animals may be approached in much the same manner. Initially, it is necessary to identify animals of potential concern, their average body weight, their daily food and water consumption, etc. Table 6.2 provides such an initial scoping of these parameters. This type of information, along with relative consumption patterns of the species of concern, are inputted into an appropriate food web model (Thomann, 1981). Such data have been used to provide estimates of contaminant intake rates for various terrestrial species in contaminant zones associated with migration of toxic substances from the RMA (i.e., offpost contamination) (Fordham and Reagan, 1991).

In order to determine if the estimated daily intake exceeds an acceptable rate (TRV), the ratio of the two values are determined and labeled a "hazard index" by HLA (1991). If this ratio exceeds one, then the extent to which an environmental media source must be reduced to achieve nonexceedance of the TRV has to be determined. The TRVs derived by HLA (1991) for the most part are inappropriately derived, as noted earlier. In addition, the term "hazard index" is inappropriate since it is used by the EPA in judging multiple chemical exposures (Calabrese, 1991).

A critical factor in terrestrial risk assessment procedures is assurance that the TRV is not exceeded. In order to achieve this goal it is necessary to have reliable information on sources of exposure, consumption patterns, and area of the animals' range in relation the above factors. For example, in the terrestrial food web of the RMA (zone 3), it is estimated that the bald eagle ingests prairie dogs

Table 6.1.   Toxicity Reference Values (TRVs) for Terrestrial Vertebrate Species of Concern Identified at Rocky Mountain Arsenal

COC	Species of Concern	Study Type	Dose (mg/kg day)	Test Species	Sensitive Test Population UF	Phylogenic UF	Trophic Level UF	Exposure Media UF	Phase 1 or 2 Metabolic Data
Arsenic	Mallard duck	LOAEL	18.9	Duck	Species-specific TRV estimated using chronic or subchronic values				
Arsenic	Great-blue heron	LOAEL	18.9	Duck	1	4	2	2	0.5
Arsenic	Mouse	LD$_{50}$	1.6	Mouse	Species-specific TRV estimated using chronic or subchronic values				
Arsenic	Prairie dog	LD$_{50}$	1.6	Mouse	1	2	2	2	0.5
Aldrin/ Dieldrin	Bald eagle	LD$_{50}$	9	Partridge	1	4	2	2	0.5
DDE/DDT	Mallard duck	LOAEL	4	Duck	Species-specific TRV estimated using chronic or subchronic values				
DDE/DDT	Great-blue heron	NOEL	0.875	Quail	0.5	4	2	2	0.5
DDE/DDT	Bald eagle	NOEL	0.875	Quail	0.5	4	2	2	0.5
DDE/DDT	Cow	LOAEL	12.1	Rat	1	4	1	1	0.5
DIMP	Mallard duck	LOAEL	410	Duck	Species-specific TRV estimated using chronic or subchronic values				
DIMP	Great-blue heron	LOAEL	60	Quail	1	4	2	2	1
DIMP	Mouse	NOAEL	300	Mouse	Species-specific TRV estimated using chronic or subchronic values				
Endrin/ Isodrin	Mallard duck	NOEL	0.3	Duck	Species-specific TRV estimated using chronic or subchronic values				
Endrin/ Isodrin	Great-blue heron	NOEL	0.3	Duck	1	4	2	2	0.5
Endrin/ Isodrin	Bald eagle	NOEL	0.3	Duck	1	4	2	2	0.5
Endrin/ Isodrin	Prairie dog	LOAEL	0.53	Mouse	1	2	2	2	0.5

continued

Table 6.1. *Continued*

COC	Toxic Intermediate	Range of Toxic Doses UF	Persistence UF	Combined UF	Modifying Factor	TRV (mg/kg day)
Arsenic	1			50	NA	3.8E−01
Arsenic	1	1	1	8	NA	2.4E+0.0
Arsenic	1	1		250	NA	6.0E−03
Arsenic	1	1	1	4	10	4.0E−02
Aldrin/ Dieldrin	4	1	10	320	50	3.0E−04
DDE/DDT	1	1	10	50	NA	8.0E−02
DDE/DDT	1	1	10	40	NA	2.2E−02
DDE/DDT	1	1	10	40	5	4.0E−03
DDE/DDT	1			20	NA	6.0E−01
DIMP	1	1	1	50	NA	8.2E+00
DIMP	1			16	NA	3.8E+00
DIMP				10	NA	3.0E+01
Endrin/ Isodrin	4			5	NA	6.0E−02
Endrin/ Isodrin	4	1	10	320	NA	1.0E−03
Endrin/ Isodrin	4	1	10	320	8	2.0E−04
Endrin/ Isodrin	4	1	10	160	NA	4.0E−03

Source: HLA, 1991.

Table 6.2.   Summary of Intakes from Zone 3 Soil, Surface Water, and Diet and Hazard Indices for the Bald Eagle

Chemical	Soil and Water (mg/kg-bw/day)	Diet (mg/kg-bw/day)	Total (mg/kg-bw/day)	Exceeds TRVs for Bald Eagle[a]	Hazard Index
Aldrin	4.1×10^{-5}	5.2×10^{-4}	5.6×10^{-4}	No	0.9
Arsenic	4.4×10^{-3}	1.2×10^{-1}	1.2×10^{-1}	No	0.25
Chlordane	3.9×10^{-4}	6.0×10^{-5}	4.5×10^{-4}	No	0.2
DDE	1.6×10^{-4}	5.2×10^{-1}	5.2×10^{-1}	Yes	130
DDT	5.2×10^{-4}	2.5×10^{-1}	2.5×10^{-1}	Yes	62
DIMP	5.7×10^{-2}	NA	5.7×10^{-2}	No	0.04
Dieldrin	1.0×10^{-3}	1.7×10^{-1}	1.7×10^{-1}	Yes	284
Endrin	5.4×10^{-3}	1.6×10^{-3}	7.0×10^{-3}	Yes	35
Fluoride[b]	6.4×10^{-1}	NA	6.4×10^{-1}	No	0.1
Sulfate[b]	1.1×10^{2}	NA	1.1×10^{2}	Yes	5.7

Source: HLA, 1991.
mg/kg-bw/day = milligrams per kilogram of body weight per day
NA = not applicable
TRV = toxicity reference value
[a]TRVs reported in Table 3.3-2 (HLA, 1991). Other raptors represented by bald eagle TRV.
[b]Cannot be addressed for dietary intake because (1) chemical is a naturally occurring component of diet or (2) cannot be modeled because of a lack of data.

(3% of its diet) and pheasants (5.4% of its diet) (Palmer and Fowler 1975). Dietary aldrin intake for the eagle was calculated to be 5.2×10^{-4} mg/kg of body weight (HLA, 1991) [(0.001 mg/kg prairie dogs $\times$ 0.03 ingestion) + (0.096 mg/kg pheasants $\times$ 0.054 ingestion) $\times$ 0.1 (percent of body weight)]. The bald eagle was also assumed to ingest 0.003 kg of soil/kg body weight per day and 0.25 liters of water/kg of body weight per day. Based on these exposure rates and given the level of contamination in zone 3 of the RMA for the various sources, Table 6.2 (HLA, Vol III, 1991) provides a summary of the exposure rates of 10 contaminants in relation to the TRV for bald eagles. The values for DDE, DDT, dieldrin, endrin, and sulfate exceed the TRVs by 130-, 62-, 284-, 35-, and 5.7-fold, respectively. Species such as the bald eagle are also a part of the aquatic food web. In the case of bald eagles, 92% of the dietary intake is from the aquatic food web (HLA, 1991). The total source of the bald eagle food in the aquatic web during winter when ice often covers the ponds/lakes is derived from the mallard duck (HLA, 1991). It should be noted that the mallard duck received 91%

of its food from consumption of algae and 9% from consumption of invertebrates.

Tables 6.3 and 6.4 provide BCFs, BMFs, and predicted tissue concentrations for the terrestrial and aquatic food webs, respectively. Table 6.4 also provides other critical model input parameters such as K2, assimilation and feeding rates, and fraction of prey in diet.

1. Why Are TRVs Derived and How Are They Used?

The TRV approach was adopted to complement the limitations of the MATissueC approach. The TRVs were considered especially useful for agents that do not bioaccumulate in food webs. In addition, the TRV methodology is designed to provide a check on the MATissueC estimates. As noted above, once the critical food web parameters are reliably estimated, it is possible to calculate a tissue level for the animal using the same concentration employed in the TRV derivation process. The newly calculated tissue concentration can be compared to the MATissueC. In this way the TRV and the MATissueC value can serve as a check on each other. Consider the following example of the bald eagle. Using the TRV approach, DDE, DDT, dieldrin, and endrin substantially exceed the acceptable rate of intake (i.e., TRV value) (Table 6.5). Using the same data from zone 3, tissue concentrations were predicted using the total BMF and soil concentration values (Table 6.6). The MATissueC value was never closely approached in these instances (i.e., DDE predicted 214 vs 3940 μg/kg MATissueC, DDT 1785 vs 3940 μg/kg MATissueC, dieldrin 118 vs 10,000 μg/kg MATissueC, and endrin 108 vs 830 μg/kg MATissueC). This large difference in acceptability estimates between the MATissueC and the TRV approach would be mitigated if the entire foraging range of the bald eagle were included in the exposure analysis, rather than only exposure from zone 3. On the other hand, the use of TRV-UFs by HLA (1991) is generally inappropriate and would underestimate the magnitude of the UF in nearly all cases.

The principal point is that these two approaches for assessing ecological risk (MATissueC and TRV) to bald eagles provide considerably divergent estimations of actual risk, with the MATissueC estimate being considerably less conservative than the TRV approach. Other applications of the TRV process for terrestrial animals at the RMA are given in Table 6.7 (mallard duck and great-blue heron) and Table 6.8 (deer mouse and prairie dog).

Table 6.3.   Bioaccumulation Factors, Biomagnification Factors, and Predicted Tissue Concentrations for the Terrestrial Food Web

BAFs for each COC[a] (Uncalibrated)

Species	Aldrin	DDE	DDT	Dieldrin	Endrin
Small bird	9.28	21.7	21.7	9.28	7.9
Small mammal	1.05	3.29	3.29	1.05	7.6
Medium mammal	0.163	3.29	3.29	1.77	7.6
Worm	6.46	2.92	2.92	6.46	29
Insect	26.4	32.2	32.2	26.4	29
Plant	0.375	0.81	2.56	0.375	0.056
Great-horned owl	5.35	2.62	2.62	5.35	7.9
American kestrel	5.35	2.62	2.62	5.35	7.9
Bald eagle	5.35	2.62	2.62	5.35	7.9

BAFs for each COC[a] (Calibrated)

Species	Aldrin	DDE	DDT	Dieldrin	Endrin
Small bird	9.28	21.7	21.7	9.28	7.9
Small mammal	1.05	3.29	3.29	1.05	7.6
Medium mammal	0.163	3.29	3.29	1.77	7.6
Worm	4	0.32	0.32	0.4	5.9
Insect	3.9	3.5	1.8	0.34	5.9
Plant	0.375	0.81	2.56	0.375	0.056
Great-horned owl	5.35	2.62	2.62	5.35	7.9
American kestrel	5.35	2.62	2.62	5.35	7.9
Bald eagle	5.35	2.62	2.62	5.35	7.9

continued

Table 6.3. *Continued*

Total BAFs for each pathway and COC (Uncalibrated)

Pathway	Aldrin	DDE	DDT	Dieldrin	Endrin	%kestrel	%owl	%eagle
s>ew>m>o,k	3.63E+01	2.52E+01	2.52E+01	3.63E+01	1.74E+03	1.02E−02	2.73E−02	
s>p>gh>m>o,k	5.56E+01	2.25E+02	7.11E+02	5.56E+01	9.75E+01	1.60E−01	4.28E−01	
s>p>gh>o,k	5.30E+01	6.83E+01	2.16E+02	5.30E+01	1.28E+01	6.30E−01	9.00E−02	
s>p>m>o,k	2.11E+00	6.98E+00	2.21E+01	2.11E+00	3.36E+00	1.60E−01	4.28E−01	
s>p>ph>e	1.86E+01	4.61E+01	1.46E+02	1.86E+01	3.49E+00			3.51E−02
s>p>pd>e	3.27E−01	6.98E+00	2.21E+01	3.55E+00	3.36E+00			2.91E−02
s>p>gh>ph>e	4.92E+02	1.48E+03	4.69E+03	4.92E+02	1.01E+02			1.78E−02
s>m						1.02E−02	2.73E−02	
s>ph								1.08E−03
s>pd								9.00E−04

BAFs for each Pathway and COC (Calibrated)

Pathway	Aldrin	DDE	DDT	Dieldrin	Endrin	%kestrel	%owl	%eagle
s>ew>m>o,k	2.25E+01	2.76E+00	2.25E+00	2.25E+00	3.54E+02	1.02E−02	2.73E−02	
s>p>gh>m>o,k	8.22E+00	2.44E+01	3.97E+01	7.16E−01	1.98E+01	1.60E−01	4.28E−01	
s>p>gh>o,k	7.82E+00	7.43E+00	1.21E+01	6.82E−01	2.61E+00	6.30E−01	9.00E−02	
s>p>m>o,k	2.11E+00	6.98E+00	2.21E+01	2.11E+00	3.36E+00	1.60E−01	4.28E−01	
s>p>ph>e	1.86E+01	4.61E+01	1.46E+02	1.86E+01	3.49E+00			3.51E−02
s>p>pd>e	3.27E+01	6.98E+00	2.21E+01	3.55E+00	3.36E+00			2.91E−02
s>p>gh>ph>e	7.26E−01	1.61E+02	2.62E+02	6.33E+00	2.06E+01			1.78E−02
s>m						1.02E−02	2.73E−02	
s>ph								1.08E−03
s>pd								9.00E−04

continued

Table 6.3. *Continued*

Total BMFs (Uncalibrated) for each Target Organism for Food Chains and Soil

Species	Aldrin	DDE	DDT	Dieldrin	Endrin
American kestrel	4.30E+01	8.04E+01	2.53E+02	4.30E+01	4.20E+01
Great-horned owl	3.05E+01	1.06E+02	3.34E+02	3.05E+01	9.20E+01
Bald eagle	9.42E+00	2.82E+01	8.92E+01	9.52E+00	2.04E+00

Total BMFs for Terrestrial Species from BAFs Calibrated with Mean RMA Soil Data and Tissue CRLs or Tissue Data[b]

Species	Aldrin	DDE	DDT	Dieldrin	Endrin
American kestrel	6.82E+00	9.76E+00	1.75E+01	9.14E+01	9.04E+00
Great-horned owl	5.76E+00	1.43E+01	2.77E+01	1.36E+00	2.00E+01
Bald eagle	1.97E+00	4.72E+00	1.04E+01	8.81E-01	6.03E+01
Prairie dog	6.42E+02	2.68E+00	8.27E+00	6.97E-01	6.41E-01
Deer mouse	1.06E+00	5.77E+00	1.12E+01	2.92E-01	2.95E+00
Ring-necked pheasant	6.93E+00	3.22E+01	6.95E+01	2.84E+00	1.31E+00

Predicted Tissue Concentrations Based on Total BMFs (Calibrated) and UCL95 Zone 3 Soil Data[c]

Species	Aldrin	DDE	DDT	Dieldrin	Endrin
American kestrel	94	443	3000	123	1619
Great-horned owl	80	648	4733	182	3586
Bald eagle	27	214	1785	118	108
Prairie dog	1	122	1414	93	115
Deer mouse	15	262	1918	39	629
Ring-necked pheasant	96	1460	11891	380	234

continued

Table 6.3. *Continued*

Predicted Tissue Concentrations Based on Total BMFs (Calibrated) and UCL95 for the Low Impact Zone Soil Data[c]

Species	Aldrin	DDE	DDT	Dieldrin	Endrin
American kestrel	18	728	2473	67	34
Great-horned owl	15	1065	3903	99	76
Bald eagle	5	352	1472	64	2
Prairie dog	0	200	1166	51	2
Deer mouse	3	430	1581	21	11
Ring-necked pheasant	18	2399	9805	207	5

(abbrev. = food web species, onpost category)
s = soil
ew = earthworm, worm
p = plant
gh = grasshopper, insect
m = deer mouse, sm mammal
ph = pheasant, sm bird
pd = prairie dog, md mammal
o = owl
k = kestrel
e = eagle

% = fraction of prey in predator diet for food pathway
BAF = bioaccumulation factor
BMF = biomagnification factor
COC = chemical of concern
RMA = Rocky Mountain Arsenal
CRL = certified reporting limit
μg/kg bw = micrograms per kilogram of body weight

Source: HLA, 1991.
[a]Ebasco, 1991. Bioaccumulation factor adjustments for earthworms and grasshoppers based on observed data.
[b]Calibrated with Offpost Operable Unit RMA data (HLA, 1991).
[c]Results in micrograms per kilogram body weight.

Table 6.4. Bioconcentration Factors, Biomagnification Factors, and Predicted Tissue Concentrations for the Aquatic Food Web

Bioconcentration Factors (Uncalibrated)[a]

Species	Aldrin	Arsenic	DDE	DDT	Dieldrin	Endrin
Small fish	8787	15.45	28704	28704	8787	4180
Large fish	8787	15.45	66761	66761	8787	4180
Invertebrate	8787	15.45	9029	9029	8787	4180
Algae	133	421.6	2973	1811	1811	96

Bioconcentration Factors (Calibrated)[b]

Species	Aldrin	Arsenic	DDE	DDT	Dieldrin	Endrin
Small fish	8787	15.45	28704	28704	500	4180
Large fish	766.7	15.45	66761	66761	1000	4180
Invertebrate	8787	27.4	9029	9029	122	4180
Algae	133	48.8	2973	1811	122	96

Other Model Input Parameters[c]

Species $k2$[a]	Aldrin	Arsenic	DDE	DDT	Dieldrin	Endrin
Small fish	0.00847	0.1287	0.0063	0.0063	0.00847	0.052
Large fish	0.00847	0.1287	0.0063	0.0063	0.00847	0.052
Mallard duck	0.0116	0.359	0.0037	0.0037	0.0116	0.035
Great-blue heron	0.0116	0.359	0.0037	0.0037	0.0116	0.035
Bald eagle	0.116	0.359	0.003096	0.003096	0.0116	0.035

continued

Table 6.4. *Continued*

Other Model Input Parameters (*continued*)

Species	Aldrin	Arsenic	DDE	DDT	Dieldrin	Endrin
Alpha[a]						
Small fish	0.9	0.8	0.787	0.787	0.9	0.85
Large fish	0.9	0.8	0.787	0.787	0.9	0.85
Mallard duck	0.9	0.8	0.9	0.9	0.9	0.9
Great-blue heron	0.9	0.8	0.9	0.9	0.9	0.9
Bald eagle	0.9	0.8	0.9	0.9	0.9	0.9
Feeding Rate		Fraction of prey in diet				
Small fish	0.019	Algae > Small fish			1	
Large fish	0.019	Algae > Large fish			1	
Great-blue heron	0.0846	Small fish > Heron			0.5	
		Invertebrate > Heron			0.5	
Mallard	0.0728	Algae > Mallard			0.91	
		Invertebrate > Mallard			0.09	
Bald eagle	0.0846	Mallard > Eagle			0.92	
Food Term	Aldrin	Arsenic	DDE	DDT	Dieldrin	Endrin
Small fish	2.02	0.12	2.37	2.37	2.02	0.31
Large fish	2.02	0.12	2.37	2.37	2.02	0.31
Great-blue heron	3.28	0.09	10.29	10.29	3.28	1.09
	3.28	0.09	10.29	10.29	3.28	1.09
Mallard duck	5.14	0.15	16.11	16.11	5.14	1.70
	0.51	0.01	1.59	1.59	0.51	0.17
Bald eagle	0.60	0.17	22.63	22.63	6.04	2.00

continued

Table 6.4. *Continued*

Pathway Specific BAFs (Uncalibrated)

Pathway	Aldrin	Arsenic	DDE	DDT	Dieldrin	Endrin
sw>a>fm>gbh	2.97E+04	6.15E+00	3.68E+05	3.40E+05	4.08E+04	4.58E+03
sw>in>gbh	3.88E+04	1.46E+00	9.29E+04	9.29E+04	2.88E+04	4.55E+03
sw>a>m>be	4.13E+02	1.08E+01	1.08E+06	6.60E+05	5.62E+04	3.27E+02
sw>in>m>be	2.70E+03	3.91E+02	3.26E+05	3.26E+05	2.70E+04	1.41E+03
sw>a>cp,buh	9.06E+03	6.52E+01	7.38E+04	7.11E+04	1.24E+04	4.21E+03

Pathway Specific BAFs (Calibrated)

Pathway	Aldrin	Arsenic	DDE	DDT	Dieldrin	Endrin
sw>a>fm>gbh	2.97E+04	2.00E+00	3.68E+05	3.40E+05	2.45E+03	4.58E+03
sw>in>gbh	2.88E+04	2.58E+00	9.29E+04	9.29E+04	4.00E+02	4.55E+03
sw>a>m>be	4.13E+02	1.25E+00	1.08E+06	6.60E+05	3.79E+03	3.27E+02
sw>in>m>be	2.70E+03	6.94E-02	3.26E+05	3.26E+05	3.75E+02	1.41E+03
sw>a>cp,buh	1.04E+03	2.12E+01	7.38E+04	7.11E+04	1.25E+03	4.21E+03

Total BMFs for Each Target Organism and Indicator Species

Species	Aldrin	Arsenic	DDE	DDT	Dieldrin	Endrin
Bald eagle	3.11E+03	1.08E+01	1.41E+06	9.86E+05	8.32E+04	1.74E+03
Great-blue heron	5.86E+04	7.61E+00	4.61E+05	4.32E+05	6.97E+04	9.13E+03
Small fish	9.06E+03	6.52E+01	3.58E+04	3.30E+04	1.24E+04	4.21E+03
Mallard duck	5.15E+03	6.25E+01	6.23E+04	4.36E+04	1.38E+04	8.68E+02
Large fish	9.06E+03	6.52E+01	7.38E+04	7.11E+04	1.24E+04	4.21E+03

continued

Table 6.4. *Continued*

Total BMF for Each Target Organism and Indicator Species (Calibrated)

Species	Aldrin	Arsenic	DDE	DDT	Dieldrin	Endrin
Bald eagle	3.11E+03	1.32E+00	1.41E+06	9.86E+05	4.16E+03	1.74E+03
Great-blue heron	5.86E+04	4.58E+00	4.61E+05	4.32E+05	2.85E+03	9.13E+03
Small fish	9.06E+03	2.12E+01	3.58E+04	3.30E+04	7.46E+02	4.21E+03
Mallard duck	5.15E+03	7.60E+00	6.23E+04	4.36E+04	6.89E+02	8.68E+02
Large fish	1.04E+03	2.12E+01	7.38E+04	7.11E+04	1.25E+03	4.21E+03

Predicted Tissue Concentrations (μg/kg) in Aquatic Species from BMF (Calibrated) and UCL95 Data for First Creek

Species	Aldrin	Arsenic	DDE	DDT	Dieldrin	Endrin
Bald eagle	0	23	125729	45744	10777	0
Great-blue heron	0	81	41107	20066	7380	0
Small fish	0	375	3190	1531	1932	0
Mallard duck	0	134	5557	2021	1784	0
Large fish	0	375	6584	3297	3227	0
Algae	0	863	265	84	316	0
Invertebrate	0	485	805	419	316	0

BAF = bioaccumulation factor
BMF = biomagnification factor
BDL = below detection limit
COC = chemical of concern
UCL95 = upper 95th confidence limit
μg/kg = micrograms per kilogram

sw = surface water
in = invertebrate
cp = carp
a = algae
fm = fathead minnow
be = bald eagle
gbh = great-blue heron
buh = bullhead fish
m = mallard duck

Source: HLA, 1991.
[a]Ebasco, June 5, 1991.
[b]Ebasco, June 5, 1991, and offpost data for arsenic and dieldrin. Other COCs not calibrated because of BDLs in surface water and biota.
[c]Ebasco 1991.

Table 6.5. Comparison of MATissueC Values Calculated from Linear One Compartment Open Model with TRV Values Calculated by HLA (1991) for Bald Eagles

COC	K2[a]	T1/2 days	Mss[b] mg/kg	Md[c] mg/kg/d	TRV[d] mg/kg/d	Md[e] TRV-HLA	Md[e] TRV-RM[f]	Recommended TRV mg
Aldrin	0.0116	59.74	9.00	0.104	0.0006	173	577	1.8
Dieldrin	0.0116	59.74	9.83	0.114	0.0006	190	633	1.8
DDE	0.003096	223.84	3.96	0.012	0.004	3	136	8.8
DDT	0.003096	223.84	3.94	0.012	0.004	3	136	8.8
Endrin	0.035	19.80	0.83	0.029	0.0002	145	966	3.0

[a]Value from Table H3 (HLA, 1991).

[b]Value from previous table (MATissueC mg/kg) (6.7).

[c]Rate of intake needed to achieve MATissueC.

[d]Value from Table 3.3.3-3. (HLA, 1991).

[e]This value means that the rate of daily intake is lower than the intake rate permitted at the MATissueC. For example, the TRV-RM method for dieldrin would permit a daily intake rate 1/633 of the rate permitted to achieve the MATissueC.

[f]Recommended Method (RM) by authors of this report.

Table 6.6. Calculated Values for MATissueC in Bald Eagles

COC	Predicted Tissue Conc. mg/kg	Ratio[a]	MATissueC mg/kg
Aldrin	0.027	0.003	9.00
Dieldrin	0.118	0.012	9.83
Endrin	0.108	0.130	0.83
DDE	0.214	0.054	3.96
DDT	1.785	0.453	3.94

$$^a\text{Ratio} = \frac{\text{Predicted Tissue Concentration}}{\text{MATissueC}}$$

2. Interconvertibility of MATissueC and TRV

The HLA (1991) report proposed the use of both the MATissueC and TRV approaches, yet it did not provide a mathematical way to estimate one value from the other. This may be accomplished by the following equation for a linear one-compartment open model (Geyer et al., 1986):

$$Md \ (mg/kg/day) = \frac{(Mss)(\ln 2)(t)}{T1/2}$$

where
   Md is the rate of daily intake needed to achieve the tissue concentration at steady state
   Mss is the tissue concentration of contaminant X at steady state
   t is the interval with which doses are administered
   T1/2 is the half-life of the COC in days

It should be recalled that:

$$T1/2 = \frac{\ln 0.5}{K2} = \frac{-0.693}{K2}$$

MATissueC can be calculated from the ratio of Predicted Tissue Concentration to MATissueC (Table 6.6):

$$\text{Ratio} = \frac{\text{Predicted Tissue Concentration}}{\text{MATissueC}}$$

Table 6.7.   Summary of Intakes from Zone 3 Soil, Surface Water, and Diet for Avian Species in the Aquatic Food Web

Chemical	Soil[b] and Water (mg/kg-bw/day)	Mallard Duck		Great-blue Heron		Hazard Index[c]	
		Diet (mg/kg-bw/day)	Total (mg/kg-bw/day)	Diet (mg/kg-bw/day)	Total (mg/kg-bw/day)	Mallard Duck	Great-blue Heron
Aldrin	4.1×10^{-5}	0	4.1×10^{-5}	0	4.1×10^{-5}	5.1×10^{-3}	1.4×10^{-2}
Arsenic	4.4×10^{-3}	8.3×10^{-2}	8.7×10^{-2}	4.3×10^{-2}	4.7×10^{-2}	2.3×10^{-1}	2.0×10^{-2}
Chlordane	3.9×10^{-4}	—	3.9×10^{-4}	—	3.9×10^{-4}	7.8×10^{-5}	4.9×10^{-2}
DDE	1.6×10^{-4}	3.1×10^{-2}	3.1×10^{-2}	2.0×10^{-1}	2.0×10^{-1}	3.9×10^{-1}	9.1
DDT	5.2×10^{-4}	1.1×10^{-2}	1.2×10^{-2}	9.8×10^{-2}	9.8×10^{-2}	1.5×10^{-1}	4.4
DIMP	5.7×10^{-2}	NA	5.7×10^{-2}	NA	5.7×10^{-2}	7.0×10^{-3}	1.5×10^{-2}
Dieldrin	1.0×10^{-3}	3.2×10^{-2}	3.3×10^{-2}	1.1×10^{-1}	5.7×10^{-2}	4.1	37
Endrin	5.4×10^{-4}	0	5.4×10^{-4}	0	5.4×10^{-4}	9.0×10^{-3}	5.4×10^{-1}
Fluoride[a]	6.4×10^{-1}	NA	6.4×10^{-1}	NA	6.4×10^{-1}	1.5×10^{-2}	2.5×10^{-2}
Sulfate[a]	1.1×10^{2}	NA	1.1×10^{2}	NA	1.1×10^{2}	1.2	1.2

Source: HLA, 1991.
— = BCFs were unavailable for lower trophic level aquatic life in the literature reviewed.
BCF = bioconcentration factor
NA = not applicable
[a]Cannot be addressed for dietary intake because (1) chemical a naturally occurring component of diet or (2) cannot be modeled because of a lack of data.
[b]Aquatic birds were conservatively assumed to ingest the same amount of soil as terrestrial birds.
[c]TRVs from Table 3.3-2, HLA, 1991.

Table 6.8. Summary of Intakes from Zone 3 Soil, Surface Water, and Diet for Mammalian Species

| | | Deer Mouse | | Prairie Dog | | Hazard Index | |
Chemical	Soil and Water[a] (mg/kg-bw/day)	Diet (mg/kg-bw/day)	Total (mg/kg-bw/day)	Diet (mg/kg-bw/day)	Total (mg/kg-bw/day)	Deer Mouse	Prairie Dog
Aldrin	4.1×10^{-5}	3.0×10^{-3}	3.0×10^{-3}	5.0×10^{-4}	5.4×10^{-4}	5.5×10^{-2}	2.0×10^{-2}
Arsenic	4.4×10^{-3}	—	4.4×10^{-3}	—	4.4×10^{-3}	7.3×10^{-1}	1.1×10^{-1}
Chlordane	3.9×10^{-4}	—	3.9×10^{-4}	—	3.9×10^{-4}	4.9×10^{-2}	1.3
DDE	1.6×10^{-4}	9.3×10^{-3}	9.4×10^{-3}	3.6×10^{-3}	3.8×10^{-3}	1.3×10^{-3}	6.2×10^{-3}
DDT	5.2×10^{-4}	3.5×10^{-2}	3.6×10^{-2}	4.2×10^{-2}	4.3×10^{-2}	5.0×10^{-2}	7.1×10^{-2}
DIMP	5.7×10^{-2}	NA	5.7×10^{-2}	NA	5.7×10^{-2}	1.9×10^{-3}	7.6×10^{-4}
Dieldrin	1.0×10^{-3}	4.7×10^{-3}	5.7×10^{-3}	4.8×10^{-3}	5.8×10^{-3}	1.1×10^{-1}	1.9×10^{-1}
Endrin	5.4×10^{-4}	5.3×10^{-2}	5.4×10^{-2}	9.7×10^{-4}	1.5×10^{-3}	2.7×10^{1}	3.7
Fluoride[b]	6.4×10^{-1}	NA	6.4×10^{-1}	NA	6.4×10^{-1}	6.8	4.7×10^{-2}
Sulfate[b]	1.1×10^{2}	NA	1.1×10^{2}	NA	1.1×10^{2}	1.4	1.2

Source: HLA, 1991 (TRVs from Table 3.3-2).

— = Arsenic was not a chemical of concern (COC) in soil (or terrestrial ecosystems); therefore, any arsenic intakes are derived from surface-water ingestion. This exposure scenario is highly unlikely for either small mammal. Chlordane BAFs for lower trophic-level life (plants, invertebrates) were unavailable in the literature reviewed.

BAF = bioconcentration factor

mg/kg-bw/day = milligrams per kilogram—body weight per day

NA = not applicable

TRV = toxicity reference value

[a]Mammals were conservatively assumed to ingest the same amount of soil and water per unit body weight as avian species.

[b]Cannot be addressed for dietary intake because (1) chemical a naturally occurring component of diet or (2) cannot be modeled because of a lack of data.

The values of the ratios given in Table 6.6 were provided in HLA-Vol. III (1991) pp. 5–33; MATissueC values were from the same reference, pp. 5–32.

Based on the above formula, a comparison was undertaken of the TRV and the calculated rate of contaminant ingestion necessary to derive the MATissueC for the bald eagle and the mallard duck with respect to the COCs. Tables 6.9 and 6.10 reveal that the rate of pollutant intake was greater to achieve the MATissueC than the TRV for each COC for both species. Of particular note is that the MATissueC methodology was substantially less conservative (i.e., less protective) than the HLA (1991) TRV process for bald eagles with respect to aldrin, dieldrin, and endrin. The divergence in estimates between the two methodologies was reduced with respect to DDE and DDT for the bald eagle, with the TRV again being more conservative, but only by a factor of 3-fold. In contrast, the TRV and MATissueC methods of HLA (1991) yielded generally similar estimates with the mallard duck.

The present analysis represents the first direct comparison of the MATissueC and TRV methodologies with respect to their capacity to predict rates of ingestion. This analysis partially supports the previous conclusion that the MATissueC methodology is less conservative than the traditional safety-factor-based approach employed within the TRV. It should be emphasized, however, that the tissue estimates derived from the HLA (1991) TRV process would be reduced considerably further if the TRV values incorporated the uncertainty factors recommended in this document (Table 4.14) in contrast to those recommended by DOD contractors at the RMA (HLA, 1991) (see last column of Tables 6.9 and 6.10). As a result of the disparity in predicted values between the two approaches (i.e., MATissueC vs TRV), it is unlikely that each can serve as a complement to the other. However, it is valuable to know that one approach is likely to be more conservative than the other. The question of which provides the most correct answer is unfortunately not addressed in this analysis, but would require a substantial bioassay research program.

In the fact of such inconsistency of predictive methods, what is the best course of action for the ecological risk assessor? The limitations of the MATissueC approach have been addressed in Chapter 3 and should be consulted again for review, if necessary. The current situation is that the MATissueC approach is likely to underestimate toxicity. Consequently, adopting the MATissueC approach

Table 6.9. Comparison of MATissueC Values Calculated from Linear One Compartment Open Model with TRV Values Calculated by HLA (1991) in Mallard Ducks

COC	$K2^a$	T1/2 days	Mss^b mg/kg	Md^c mg/kg/d	TRV^d mg/kg/d	Md^e TRV-HLA	Md^e TRV-RM^f	Recommended mg/kg/day
Aldrin	0.0116	59.74	10	0.116	0.008	14.5	295	0.0004
Dieldrin	0.0116	59.74	10	0.116	0.008	14.5	295	0.0004
DDE	0.0037	187.30	3.94	0.014	0.08	.175	3.5	0.004
DDT	0.0037	187.30	3.94	0.014	0.08	.175	3.5	0.004
Endrin	0.035	19.80	0.83	0.029	0.06	.48	9.6	0.003

Source: HLA, 1991.

[a]Value from Table H3 (HLA, 1991).

[b]Value from HLA, 1991 III-5-32.

[c]Rate of intake needed to achieve MATissueC.

[d]Value from Table 3.3.3-3. (HLA, 1991).

[e]This value means that the rate of daily intake is different by a factor of the number given below than the intake rate permitted at the MATissueC. For example, the TRV-RM method for dieldrin would permit a daily intake rate 1/295 of the rate permitted to achieve the MATissueC.

[f]Recommended Method (RM) by authors of this report.

Table 6.10. Derivation of TRVs with Recommended Methodology (RM) for Bald Eagle (Endangered Species) and Mallard Duck (Nonendangered Species) for Selected COCs:[a] A Comparison with (1) DOD-Derived TRVs (i.e., HLA)[b] and (2) MATissueC Dose Rate Values

BALD EAGLE
Example 1

COC	Species of Concern	Study Type	Dose mg/kg/d	Test Species
Aldrin/ Dieldrin	Bald Eagle	LD_{50}	9	Partridge

UF Interspecies	UF Intraspecies	LD_{50} NOAEL	Total UF
100	10	50	50,000

TRV Dose/UF (9/50,000)	$\dfrac{\text{Md}}{\text{TRV-RM}}$	$\dfrac{\text{Md}}{\text{TRV-HLA}}$
1.8×10^{-4} (.00018)	577 (aldrin) 633 (dieldrin)	173 (aldrin) 190 (dieldrin)

ALDRIN/DIELDRIN
UF Decision Scheme

	UF Size
Interspecies UF	
Phylogenetic Relatedness: Orders within Class	100
Bald Eagle: Order Falconiformes	
Partridge: Order Galliformes	
Intraspecies UF	
Endangered Species	10
Acute-to-Chronic UF	
LD_{50} to Chronic	50
Total UF	50,000

continued

Table 6.10. *Continued*

BALD EAGLE
Example 2

COC	Species of Concern	Study Type	Dose mg/kg/d	Test Species
DDE/DDT	Bald Eagle	NOEL	.875	Quail

UF Interspecies	UF Subchronic Less-Than-Lifetime	Intraspecies	Total UF
100	10	10	10,000

TRV (mg/kg/day) Dose/UF (.875/10,000)	Md TRV-RM	Md TRV-HLA
.000088 (8.8×10^{-5})	136 (DDE/DDT)	3 (DDE/DDT)

DDE/DDT
UF Decision Scheme

Interspecies UF   **UF Size**
Phylogenetic Relatedness: Orders within a Class.................   100
Bald Eagle: Order Falconiformes
Quail: Order Galliformes

Intraspecies UF
Endangered species   10

Less-than-Lifetime UF
Less than 25% of Lifespan   10

Total UF 10,000

continued

Table 6.10. *Continued*

BALD EAGLE
Example 3

COC	Species of Concern	Study Type	Dose mg/kg/d	Test Species
Endrin	Bald Eagle	NOEL	0.3	Duck

UF Interspecies	UF Subchronic Less-Than-Lifetime	Intraspecies	Total UF
100	10	10	10,000

TRV (mg/kg/day) Dose/UF (0.3/10,000)	$\dfrac{\text{Md}}{\text{TRV-RM}}$	$\dfrac{\text{Md}}{\text{TRV-HLA}}$
3×10^{-5} (0.0003)	966	145

ENDRIN
UF Decision Scheme

Interspecies UF   **UF Size**
  Phylogenetic Relatedness: Order within a Class   100
  Bald Eagle: Order Falconiformes
  Duck: Order Anseriformes

Intraspecies UF
  Endangered species   10

Less-than-Lifetime UF
  Less than 25% of Lifespan   10

Total UF 10,000

continued

Table 6.10. *Continued*

MALLARD DUCK
Example 1

COC	Species of Concern	Study Type	Dose mg/kg/d	Test Species
Endrin	Mallard Duck	NOEL	0.3	Duck

UF Less-Than-Lifetime	UF Life Stage	Total UF
10	10	100

TRV (mg/kg/day) Dose/UF (0.3/100)	Md TRV-RM	Md TRV-HLA
3×10^{-3} (0.003)	9.6	0.48

ENDRIN
UF Decision Scheme

	UF Size
Interspecies UF	
Same Species..	1
Intraspecies UF	10
Less-than-Lifetime UF	
Less than 15% of Lifespan..	10
Total UF..	100

continued

Table 6.10. *Continued*

MALLARD DUCK
Example 2

COC	Species of Concern	Study Type	Dose mg/kg/d	Test Species
DDE/DDT	Mallard Duck	LOAEL	4	Duck

UF LOAEL to NOAEL	Less-Than-Lifetime	Intraspecies	Total UF
10	10	10	1000

TRV (mg/kg/day) Dose/UF (4/1,000)	$\dfrac{Md}{TRV\text{-}RM}$	$\dfrac{Md}{TRV\text{-}HLA}$
4×10^{-3} (0.004)	3.5	0.175

DDE/DDT
UF Decision Scheme

	UF Size
Interspecies UF	
Same Species	1
Intraspecies UF	10
Less-than-Lifetime UF	
Less than 15% of Lifespan	10
LOAEL to NOAEL	10
Total UF	1,000

continued

Table 6.10. *Continued*

MALLARD DUCK
Example 3

COC	Species of Concern	Study Type	Dose mg/kg/d	Test Species
Aldrin/ Dieldrin	Mallard Duck	LOAEL	0.4	Duck

UF LOAEL to NOAEL	Less-Than-Lifetime		Intraspecies	Total UF
10	10		10	1,000

TRV (mg/kg/day) Dose/UF (0.4/1,000)	Md TRV-RM	Md TRV-HLA
4×10^{-4} (0.0004)	295	14.5

ALDRIN/DIELDRIN
UF Decision Scheme

	UF Size
Interspecies UF Same Species	1
Intraspecies UF	10
Less-than-Lifetime UF Less than 15% of Lifespan	10
LOAEL to NOAEL	10
Total UF	1,000

[a]Followed recommended procedures of Table 4.14 for UF selection.
[b]HLA TRVs for the bald eagle and mallard duck are given in Tables 6.9 and 6.10, respectively.

especially when widely divergent estimates exist, is unwise. Thus, it would appear that the more conservative of the two approaches should be adopted until more data are available to offer improved extrapolative and risk assessment insights.

7

Sediment Quality Criteria

A. INTRODUCTION

Ecological risk assessment procedures have had a principal focus on establishing ambient water quality criteria. Despite the establishment of nationwide criteria and their achievement in many areas, it has been argued that such environmental criteria may not ensure protection of organisms associated with the sediments. That this should be the case is not surprising since water quality criteria were not designed to protect sediment-associated organisms. However, the increased awareness of both the restricted application of water quality criteria and the occurrence of sediment degradation in the presence of acceptable water quality values has led to the suggestion that sediment quality criteria (SQC) are a necessary complement to water quality criteria (WQC). Moreover, Chapman (1989) has articulated five reasons to support the derivation of SQC:

- Various toxic contaminants present in only trace concentrations in the water column accumulate in sediment to high concentrations;
- Sediment may act as both a reservoir and a source of contaminants to the water column;
- Sediments accumulate contaminants over time, while the water column contaminant levels are more variable;
- Sediment contaminants, as well as toxic substances in the water column, affect benthic and other sediment-associated organisms;
- Sediments are crucial components of the aquatic habitat for numerous aquatic biota.

Given this collective perspective supporting the need to develop SQC, it is not surprising that various research teams have considered various methodological approaches by which SQC could be derived. For example, in 1986, the seventh annual meeting of the Society of Environmental Toxicology and Chemistry (SETAC) organized a session on SQC Development in Support of Environmental Protection in which four independent methodologies were presented (Chapman et al. 1987). In a subsequent paper Chapman (1989) offered a critical analysis of the principal emerging methodologies (Table 7.1). The subsequent descriptions of SQC will be based on the review of Chapman (1989), as well as the original reports, as noted in their respective sections.

B. APPARENT EFFECTS THRESHOLD APPROACH (Tetra Tech, 1985, 1986)

1. Principle

The apparent effects threshold (AET) approach uses field data on chemical concentrations in sediments and at least one indicator of bioavailability/bioeffects (e.g., sediment bioassays, benthic infaunal community structure, bottom-fish histopathological abnormalities, bioaccumulation) to assess the concentration of a specific contaminant above which statistically significant biological effects are compared to reference (i.e., control) sediments.

2. Operational Aspects

1. The AET approach emphasizes the use of site-specific biological effect indicators (e.g., benthos, sediment bioassays) over area-specific indicators (e.g., bottom fish histopathology) or noneffects based indicators (e.g., bioaccumulation).
2. This approach usually is based on dry-weight normalized chemical concentrations rather than normalization to other variables, such as total organic carbon. Such normalization procedures have been shown to be the most appropriate.

3. Advantages

1. The AET approach offers wide flexibility since it is not constrained by the nature of the contaminant as long as biological effects

Table 7.1.   Approaches Reviewed for Establishing Sediment Quality Values

Approach	Concept
Reference Approach	Sediment quality values are based on chemical concentrations in a pristine area or an area with acceptably low levels of contamination.
Water Quality Criteria Approach	Contaminant concentrations in interstitial water are measured directly and compared with U.S. EPA water quality criteria.
Equilibrium Partitioning (Sediment-Water) Approach	A theoretical model is used to describe the equilibrium partitioning of a contaminant between sedimentary organic matter and interstitial water. A sediment quality value for a given contaminant is the organic carbon normalized concentration that would correspond to an interstitial water concentration equivalent to the U.S. EPA water quality criterion for the contaminant.
Equilibrium Partitioning (Sediment-Biota) Approach	Acceptable contaminant body burdens for benthic organisms are based on existing regulatory limits. Sedimentary contaminant concentrations that would correspond to these body burdens under thermodynamic equilibrium are established as sediment quality values.
Field Bioassay Approach	Relationships between chemical concentrations and biological responses are established by exposing test organisms to field-collected sediments with measured contaminant concentrations.
Screening Level Concentration (SLC) Approach	The SLC approach estimates the sediment concentration of a contaminant above which less than 95 percent of the total enumerated species of benthic infauna are present. SLC values are empirically derived from paired field data for sediment chemistry and species-specific benthic infaunal abundances.
Apparent Effects Threshold (AET) Approach	An AET is the sediment concentration of a contaminant above which statistically significant biological effects (e.g., amphipod mortality in bioassays, depressions in the abundance of benthic infauna) would always be expected. AET are empirically derived from paired field data for sediment chemistry and a range of biological effects indicators.

continued

Table 7.1. *Continued*

Spiked Bioassay Approach	Dose-response relationships are established by exposing test organisms to sediments that have been spiked with known amounts of chemicals or mixtures of chemicals. Sediment quality values are determined for sediment bioassays in the manner that aqueous bioassays were used to establish U.S. EPA water quality criteria.

Source: Chapman, 1989.

    can be statistically assessed in comparison to a control.

2. No assumptions are required for the specific mechanisms of interactions between organisms and toxic contaminants.
3. This approach provides sediment quality values based on non-contradictory evidence of biological effects.
4. The AET clearly outperformed the equilibrium partitioning (EP) method in ability to correctly identify impacted and severely impacted sediments at Puget Sound (Tetra Tech, 1986). A comparative uncertainty analysis also indicated that the AET method was accompanied with much less uncertainty than the EP approach.

4. Disadvantages

1. The AET approach does not establish cause-and-effect relationships; it only identifies associational relationships between contaminants and biological effects.
2. Results may be modulated by the presence of unmeasured toxic contaminants.
3. The AET values may not reflect chronic responses.
4. No procedure exists to separate single contaminant effects from the effects of all contaminants combined.
5. This approach may both under- and overestimate adverse effects.
6. A protocol is required that assures representative sampling.

5. Present Applications

1. AET values have been used for 64 organic and inorganic toxic chemicals in Puget Sound (Tetra Tech, 1986).
2. The AET approach could be used for the Great Lakes using National Oceanic and Atmospheric Administration (NOAA) National Status and Trends Program.
3. Biological effects currently used as part of AET include three different sediment bioassays (amphipod, 10-day test; oyster larvae, 2-day test; bacterial luminescence, mictrotox test).

6. How a Calculation is Made

Figures 7.1 and 7.2 illustrate how the AET approach is carried out using data from the Puget Sound study (Tetra Tech, 1986). By use of bars in the figures, three subpopulations of all sediments assessed for chemical levels and biological effects are shown. The figures provide information on sediments that (1) did not show significant faunal decrements, (2) did not show significant toxicity [as measured by amphipod bioassays (i.e., mortality), oyster larvae bioassays (i.e., abnormality), benthic infaunal abundances, bacterial luminescence bioassays (Microtox bioassay)], and (3) showed either toxicity or infaunal depression. The horizontal axis provides sedimentary levels of either 4-methylphenol (Figure 7.1) or lead (Figure 7.2) on a log scale. The AET as derived by Tetra Tech (1986) was based on lead and 4-methylphenol concentration ranges associated with sediments that do not display statistically significant biological effects.

The figures also indicate the term *potential effect threshold*. This term represents the contaminant concentration below which no statistically significant biological effects were seen. However, some degree of toxicity or benthic effects are found at some (but not all) of the stations with higher concentrations. Because of the uncertainties associated with establishing the precise causal nature of the effects, Tetra Tech (1986) recommends that SQC not be established at the *potential effect threshold*. However, *the apparent (as compared to potential) benthic effects* and toxicity thresholds are levels above which all samples display infaunal depressions or toxicity, respectively.

Tetra Tech (1986) emphasized such figures need to be interpreted with professional judgment. For instance, sediment station SP-14 (Figure 7.1) showed marked toxicity and depressed infaunal abundances which they believed were related to high 4-methyl phenol concentrations (7,400-fold higher than reference values). In contrast, the same sediment from SP-14 had low levels of lead that were not critical in deriving the AET for lead. Even though the SP-14 with low lead (and high 4-methyl phenol) caused toxicity, other sediments with higher lead values showed no significant biological effects. This led the investigators to conclude that the effects at SP-14 were not caused by lead, but more probably were related to 4-methyl phenol or other covarying substance.

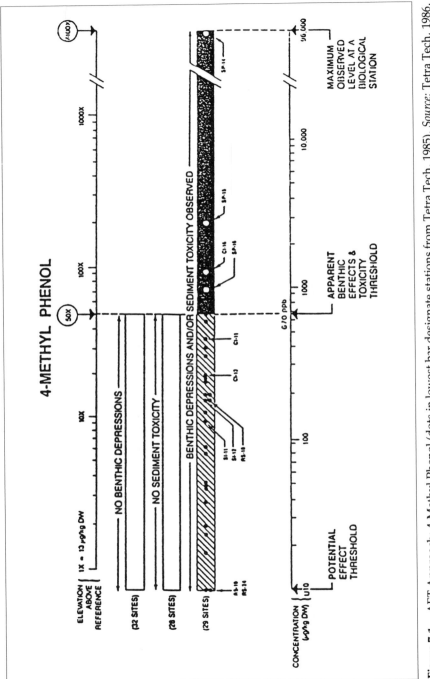

Figure 7.1.   AET Approach: 4-Methyl Phenol (dots in lowest bar designate stations from Tetra Tech, 1985). *Source:* Tetra Tech, 1986.

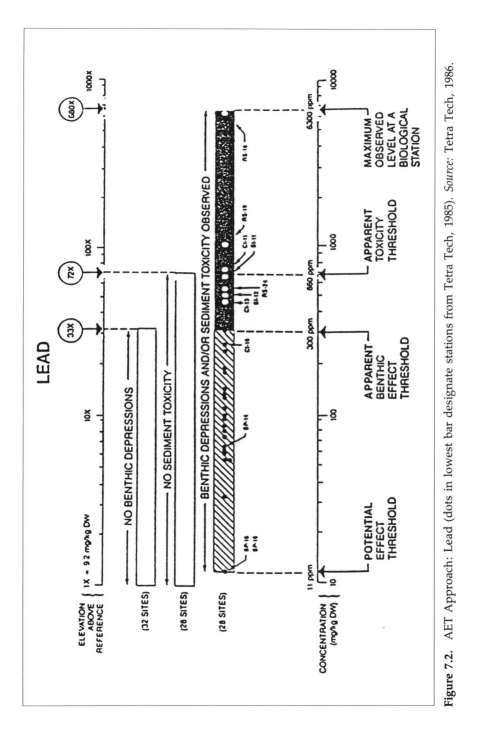

Figure 7.2. AET Approach: Lead (dots in lowest bar designate stations from Tetra Tech, 1985). *Source:* Tetra Tech, 1986.

C. SEDIMENT/WATER EQUILIBRIUM PARTITIONING (EP) APPROACH

1. Principle

The EP approach describes equilibrium partitioning of nonpolar, nonionic organic contaminants between sedimentary organic matter and interstitial water (Pavlou, 1987; DiTorro et al., 1991; Shea, 1988). More specifically, the distribution coefficients for the individual nonpolar organic contaminants are employed to establish chemical concentrations in sediments which at equilibrium will yield concentrations in the interstitial water that are equivalent to the water quality criteria. The interstitial water concentrations are estimated from organic carbon normalized partition coefficients.

2. Advantages

1. EP methodology can be linked with the EPA water quality criteria.
2. The EP method addresses the issue of bioavailability.
3. The EP method is broadly applicable, since data on equilibrium partitioning are available for numerous agents.

3. Disadvantages

1. Criteria cannot be developed for agents unless water quality criteria exist for them. Only a limited number of agents have such criteria.
2. Large variability in measurement of some partition coefficients exists (Kenaga and Goring, 1980; Rapaport and Eisenreich, 1984). In some instances, Kow values for the same compound may vary over an order of magnitude. According to Tetra Tech (1986), such lack of agreement in reported Kow values may be caused at least in part by the use of different techniques for Kow determination (e.g., shake flasks, generator columns, reverse-phase HPLC, and approximations based on fragment constants). Another factor that enhances the uncertainty of predictions of interstitial water contaminant concentration based on Kow relationships is that colloidal or dissolved organic matter in interstitial water has the capacity to affect variation from KD values predicted from studies with a pure aqueous phase. Thus, the presence of such dissolved organic matter in the interstitial water could enhance the solubility of lipophilic agents (e.g., DDT), resulting in reducing actual sediment-water-partition coefficients (Tetra Tech 1986).

3. The EP method is most valid for nonpolar organic agents. Validity for other agents is much less compelling.
4. The EP Method does not effectively deal with the issue of interactions.
5. The EP approach does not use toxicological data from the sediment of concern.
6. Evaluation of the EP method at Puget Sound by Tetra Tech (1986) revealed that this method correctly identified less than 50% of the impacted and even severely impacted stations. Confidence in the use of the EP approach was also challenged in the Tetra Tech (1986) report by the results of an uncertainty analysis. This assessment indicated that the uncertainty of the EP method was extraordinarily high, ranging from less than one to up to six orders of magnitude of the calculated values. This was attributed to uncertainty in the estimation of theoretical constants and in the applicability of water quality criteria employed in the method itself.

4. How Calculations Are Made

In the EP methodology, an organic carbon-normalized SQC is expressed in units of μg chemical/g organic carbon in sediment (Figure 7.3). This value is derived from the WQC and the partition coefficient for the chemical. The SQCs are site-specific, since they employ the organic carbon content of the sediments at the particular site.

The next step is the estimation of environmental stress via the development of a toxicity quotient (TQ). The TQ is defined as the ratio of the measured concentration of a chemical in sediment to the SQC for the specific agent. A TQ greater than one suggests stress to aquatic fauna. TQs have the capacity to both over- and underestimate stress to aquatic species, since the WQC takes bioaccumulation into account even if this is not a concern for sediment organisms. On the other hand, the TQ approach ignores the possibility of additive and/or synergistic effects of multiple agents (Burmaster et al. 1991).

As implied above, the EP approach assumes that toxicity of the agent to benthic organisms is related to the interstitial (pore) water concentration and not directly to the total concentration of the agent in the sediment. In addition, the EP method assumes that the interstitial water concentration of the agent is determined by partitioning between the sediment and the water at concentrations less

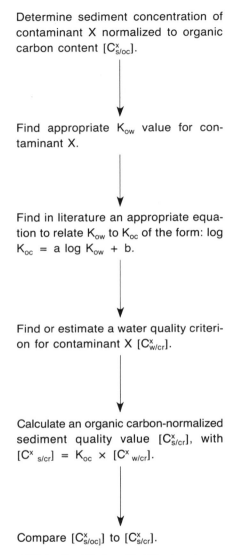

Determine sediment concentration of contaminant X normalized to organic carbon content $[C^x_{s/oc}]$.

Find appropriate K_{ow} value for contaminant X.

Find in literature an appropriate equation to relate K_{ow} to K_{oc} of the form: log K_{oc} = a log K_{ow} + b.

Find or estimate a water quality criterion for contaminant X $[C^x_{w/cr}]$.

Calculate an organic carbon-normalized sediment quality value $[C^x_{s/cr}]$, with $[C^x_{s/cr}]$ = K_{oc} × $[C^x_{w/cr}]$.

Compare $[C^x_{s/oc}]$ to $[C^x_{s/cr}]$.

Figure 7.3. Sediment-Water Equilibrium Partitioning Approach. See text for definition of terms. *Source:* Burmaster et al., 1991.

than saturation in both phases. Since the partitioning of nonpolar hydrophobic organic contaminants is controlled by the amount of organic carbon in the sediment, the higher the organic carbon content, the less partitioning of the agent occurs to the water phase (Burmaster et al., 1991). These relationships may be described by:

$$Csed/Cw = foc \times Koc \qquad (7.1)$$

Csed = concentration in sediment (mg/kg)
Cw   = interstitial water concentration (mg/L)
foc = fraction of organic carbon in sediment (dimensionless)
Koc = water/carbon partition coefficient (dimensionless)

The EPA obtains the Koc from the octanol/water partition coefficient via the equation:

$$\log Koc = 0.00028 + (0.983 \log Kow)$$

The EPA calculates the SQC as that sediment concentration at equilibrium with an interstitial water concentration equal to the WQA. This is accomplished by substituting a chronic WQC for the interstitial water concentration in Equation 7.1. The agency estimates SQC values using the mean and 95% CI of log Kow to indicate the extent of uncertainty associated with the criteria. In 1988, the EPA developed interim SQC for 13 nonpolar hydrophobic organic contaminants (e.g., chlorinated hydrocarbon pesticides DDT, dieldrin, endrin, PAH-phenanthrene, fluoranthene, phyrene, benzo(a)pyrene, benzo(a)anthracene)(DiTorro et al. 1991).

In order to calculate a SQC value, the lowest concentration in water (μg/L) with a chronic effect is determined (e.g., phenanthrene has a FCV of 6.3 μg/L). This value is then inserted into Equation 7.1 after the foc and Koc have been determined. The estimation of the interim SQC for phenanthrene is 138 μg/g carbon (33 to 605 for the lower and upper 95% CI).

5. Applications

Bioaccumulation of organic contaminants in fish can be estimated from concentrations in stream sediments using equilibrium (Burmaster et al., 1991; Thomann et al., 1992) and fugacity models (Burmaster et al., 1991; Mackay and Paterson, 1982). The equilibrium model (Equation 7.1) assumes a closed system with no dilution by flowing water. This will result in concentrations in fish that are in equilibrium with those living in interstitial water. Therefore, the concentration of a contaminant in fish is contingent on its concentration in pore water and on a BCF, which is quantitatively predicted from its Kow:

$$Cfish/Cw = BCF = 0.048 \times Kow$$

Cfish = concentration of contaminant in fish (mg/kg)
Cw = concentration in interstitial water (mg/L)
BCF = L/kg
Kow = octanol/water partition coefficient (L/kg)

With respect to the fugacity model, pore water concentrations are in equilibrium with sediment concentrations and concentrations in fish are in equilibrium with concentrations in water. However, in contrast to the EP model, the fugacity model assumes an open system in which flow dilutes concentrations in the water column compared to those in pore water (Mackay and Paterson 1982). Concentration of the agent(s) in column water and biota can be made by employing the following values: (1) pore water concentrations in sediments estimated from sediment concentration; (2) rate of groundwater discharge (if applicable); (3) several physical and chemical properties of the agents of concern including Henry's law constant, vapor pressure, and log Kow; (4) emission rates for all compartments; (5) environmental temperature; (6) vector flow of the water compartment; and (7) an assumed (or measured) upstream (or background) concentration for each compound.

Limited data at present indicate that the EP model may overestimate measured concentrations, while the fugacity model may underestimate actual tissue levels (Burmaster et al. 1991; Connolly and Pedersen 1988).

D. CHEMICAL-BY-CHEMICAL AND CHEMICAL MIXTURE CRITERIA

1. Sediment Bioassay Approach (Chapman, 1986, 1989)

Sediment bioassays and chemical analysis would be conducted on field-obtained sediments of concern and with control (i.e., reference) sediments.

a. Assumption

The laboratory assessment is an accurate predictor of in situ biological effects.

b. Advantages

It follows the scheme employed to derive water quality criteria. By assessing a complex mixture, it provides an empirical way to consider the issue of mixtures.

c. Disadvantages

No insight is provided on chemical-specific sediment quality values. The results of findings can be modulated by the presence of unmeasured, covarying toxic contaminants. Methodological approaches for conducting sediment bioassays require further development and optimization. Issues concerning how to select representative organisms, and incorporating diverse feeding types, sediment types, exposure, and duration require refinement.

2. Screening Level Concentration (SLC) Approach (Neff et al. 1987)

a. Principle

The SLC approach is a field-based approach that estimates the highest concentration of a particular toxicant (i.e., nonpolar organic contaminant) in sediment that can be tolerated by 95% of benthic infauna species for that sediment. The SLC is defined by Neff et al. (1988) as the concentration of a nonpolar organic contaminant in sediment that, if exceeded, could lead to environmental degradation and, therefore, warrant further investigation.

SLCs are calculated from contaminant concentrations which have been normalized to sediment organic carbon concentrations instead of contaminant concentrations in bulk sediment (Neff et al., 1988). This is based on the theory of equilibrium partitioning in which the bioavailability of nonpolar organic contaminants in sediment is a function of the organic carbon content of the sediment, the lipid concentration of the organism, and the relative affinities of the agent for sediment organic carbon as compared to organism lipid. The linkage of the equilibrium partitioning with bioavailability of nonpolar organic pollutants within a toxicological context is seen in the study of Abernethy et al. (1988), which reported that the acute toxicities of 38 hydrocarbons and chlorinated hydrocarbons to fresh and saltwater microcrustaceans are predicted in large part by organism-water partitioning.

The SLC is derived as follows: the presence of selected benthic indicator species are matched with the total bulk sediment concentrations of particular individual contaminants to estimate the level of each toxicant at which 90% of the particular individual species is present (i.e., survived). The concentration is referred to as the Species Screening Level Concentration (SSLC). SSLCs are estimated for all appropriate species. The SSLCs for all species are compared with sediment concentrations of each measured contaminant to determine the level around which 95% of the SSLCs occur. The final concentration is called the SLC.

b. Data Requirements

The minimum data needed to estimate SLCs for a contaminant are: 20 stations for each SSLC, 20 taxa for each SLC, a gradient of contamination, and a similar level of taxonomy (i.e., similar species if possible).

c. Advantages

The SLC can be employed with any chemical toxicant. It is consistent with U.S. EPA Water Quality Criteria goals. No toxicological mechanisms need to be assumed. It can make use of existing databases and methods. Since this procedure employs field data, there is no need to extrapolate from laboratory to field conditions.

d. Disadvantages

Field collection activities may be challenging, since it requires infaunal taxonomic identification. This approach is markedly influenced by the range and distribution of contaminant concentrations and the particular species used to produce them. Criteria for the selection of representative sensitive species are not yet established. The SLC suffers from an inability to separate out single contaminant effects from the total effects of an entire mixture.

The SLC approach will have the capacity to be unnecessarily overprotective if all concentrations observed are below levels in sediments that would cause adverse effects in the benthic infauna. Conversely, if data were principally from an extensively contaminated area, most of the pollutant-sensitive species would be absent and the SLC would be derived principally on pollutant-tolerant species, thereby leading to a SLC that would be too high.

e. Applications

Neff et al. (1988) contend that the SLC approach cannot be employed alone to determine whether a particular sediment is hazardous, because the sediment may contain agents for which no SLC exists. They indicate that the SLC method could be used in conjunction with the Triad approach (see Section E, this chapter) for verification purposes. The SLC approach could be strengthened by the use of a wide range and distribution of contaminant concentration so that valid a posteriori inferences can be included.

SLCs have been established for nine chemical agents. With respect to freshwater sediments, SLCs have been calculated for total PCBs, DDT, dieldrin, chlordane, and heptachlor epoxide. Nine SLCs have been calculated for marine sediments. These include total PCBs, DDT, naphthalene, phenanthrene, fluoranthene, benz(a)anthracene, chrysene, pyrene, and benzo(a)pyrene.

f. Example of SLC Calculation

1. Data Requirements
    a. Observations of species composition of benthic infauna are necessary.
    b. Concentration of the organic contaminant to be assessed in the sediment is needed.
    c. Concentration of total organic carbon in the sediments is required.
    d. At least 20 observations of the presence of a particular species in sediments with different concentrations of the contaminant of interest are needed. This will be necessary for the calculation of an SSLC.
    e. At least 10 SSLCs are necessary to calculate an SLC.
2. Computational Procedures
    a. Contaminant concentrations are normalized to the total organic carbon concentration of the sediment at each station with the following equation:

$$\text{TOC} - \text{normalized contaminant concentration} =$$
$$\text{X/TOC} \; (\mu g \text{ contaminant/g organic carbon})$$

where
   X is the contaminant concentration in the bulk sediment (μg contaminant/kg sediment dry weight).
   TOC is the concentration of total organic carbon in the sediment (g organic carbon/kg sediment dry weight).
b. The organic carbon normalized concentration of the chemical in the sediment for all samples in which the species was present is plotted against the cumulative frequencies of such co-occurrences, going from the least to the most contaminated station.
c. Based on the plot, the sediment contaminant concentration below which 90% of the samples containing the species occurred is estimated. This concentration is the SSLC.
d. This procedure is repeated for each benthic species at 20 or more stations and will produce multiple SSLCs for a specific contaminant.
e. Rank order all the SSLC values.
f. Calculate the five percentile of the SSLC distribution by linear interpolation. The five percentile value is termed the SLC.
g. See Figure 7.4 for a schematic representation.

E. SEDIMENT QUALITY TRIAD (SQT) APPROACH (Chapman 1986; Long and Chapman 1985; Chapman et al. 1986, 1987)

1. Principle

The SQT methodology relies on the interrelationships between sediment chemistry to estimate contamination, sediment bioassays to determine toxicity, and in situ bioeffects to determine alteration of resident communities. Chemistry and bioassay data are derived from laboratory measurement of field-collected sediment. In situ studies may involve various benthic community assessments.

The AET approach and the SQT have similar data requirements. However, the SQT requires the defining of "minimal" and "severe" biological effects in order to derive criteria in terms of chemical concentrations below which biological effects would be predicted to be minimal and above which significant adverse effects may be expected.

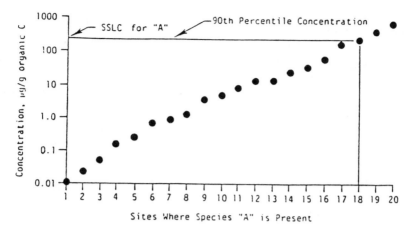

a. Calculation of Species Screening Level Concentration (SSLC)

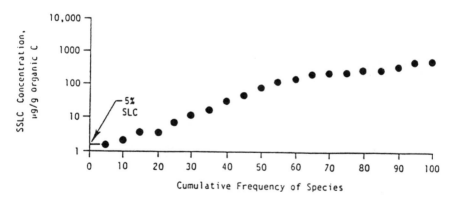

b. Calculation of Screening Level Concentration (SLC)

Figure 7.4.   Calculation of Screening Level Concentration (SLC) for SLC Approach (from Tetra Tech, 1986)

2. Advantages

The use of three independent measures permits the differentiation of toxicity related to contamination from natural variability and to laboratory artifacts. The approach is also not restricted to acute or chronic effects.

3. Disadvantages

No statistical criteria have been developed for use with the triad. It also suffers from a lack of adequate standardization of sediment bioassays testing and an inability to effectively deal with the mixture issue.

4. Applications

SQCs using the triad approach have been derived for lead, PAHs and PCBs (Chapman, 1986).

5. Example of a Calculation (Based on Chapman 1986 at Puget Sound)

a. Data

1. Sediment Chemistry
   Spatial distribution of chemicals of concern (Pb, total PCBs, and combustion PAHs).
2. Sediment Bioassays
   Three types of sediment bioassays have been reported.
   • Acute lethality test in amphipod.
   • Respiration effects test in oligochaete.
   • Fish cell anaphase aberration test.
     Each bioassay test was equally weighted. Each bioassay involves testing of the model with the various sediments for which contaminant concentrations have been determined.
3. In Situ Studies
   • Frequency of selected liver lesions in English sole from different areas of Puget Sound was recorded.
   • Frequency of liver lesions was determined for each embayment.
4. Analysis

Determine whether biological effects increase with increasing concentrations of sediment chemicals. Tables 7.2 and 7.3 provide a comparison of biological effects frequencies with sediment concentrations of selected chemical contaminants for Puget Sound.

Table 7.2. Comparison of Biological Effects Frequencies with Sediment Concentrations of Selected Chemical Contaminants for Puget Sound

Area	Effects Frequency (%)[a]	Chemical Contaminants[b]		
		Pb (μg/g)	CPAHs (μg/g)	Total PCBs (μg/g)
Sediment bioassays				
Samish Bay	15	10	0.8	ND
Bellingham Bay	20	20	1.2	ND
Everett (Port Gardner)	35	30	2.0	ND
Commencement Bay	35	30	0.8	0.10
Elliott Bay	45	50	5.0	0.20
Case Inlet	50	20	0.2	0.01
Sinclair Inlet	55	90	3.8	0.10
Commencement Bay Waterways	75	130	6.8	0.90
Elliott Bay Waterfront and Duwamish River	80	800	24.0	0.80
Bottom fish histopathology				
Case Inlet	0	20	0.2	0.01
Commencement Bay	0–5	30	0.8	0.10
Elliott Bay	0–3	50	5.0	0.20
Sinclair Inlet	2–4	90	3.8	0.10
Commencement Bay Waterways	6–40	130	6.8	0.90
Elliott Bay Waterfront and Duwamish River	8–26	800	24.0	0.80

Source: P. M. Chapman, 1986.
ND, not detected; detection limits = 0.005 μg/g. CPAHs, combustion polyaromatic hydrocarbons; PCBs, polychlorinated biphenyls.
[a]For sediment bioassays, the frequency equals the number of stations showing toxicity divided by number of stations tested, rounded to the nearest 5%; for fish histopathology, the frequency equals the ranges for selected liver lesions.
[b]Mean dry weight values rounded to the nearest factor of 10.

Table 7.3. Summary Comparison of Biological Effects Frequencies with Sediment Concentrations of Selected Chemical Contaminants

Effects Frequency (%)	Chemical Contaminants		
	Pb (μg/g)	CPAHs (μg/g)	Total PCBs (μg/g)
Sediment bioassays			
15–15	20–50	0.20–5.0	0.01–0.10
55–80	90–800	3.8–24.0	0.10–0.90
Bottom fish histopathology			
0–5	20–90	0.2–3.8	0.01–0.20
6–40	130–800	6.8–24.0	0.80–0.90

Source: P. M. Chapman, 1986.

These data were then used to derive SQC for the three COCs as presented in Table 7.4.

F. COMPARISON OF SQC APPROACHES WITH MATissueC APPROACH

According to Chapman (1989), body burden criteria (e.g., MATissueC methodology) are of limited value with respect to SQC since bioaccumulation of sediment-associated contaminants by aquatic organisms is not a very practical estimate of adverse ecological impact. This is due principally to a very meager database establishing residue-effect relationships. The MATissueC approach is practically useful for the agents that tend to biomagnify. However, given the relatively small number of agents for which this is a concern and the fact that it is most evident at higher trophic levels, the SQC can essentially proceed with current approaches without major concern over the biomagnification issue.

Despite the above comments concerning the general trend of SQC methods of not addressing bioaccumulation and biomagnification phenomena, the AET methodology is strikingly similar to the MATissueC methodology with its emphasis on field data and bioavailability and/or bioeffects. Both approaches are essentially ecological cross-sectional evaluations (i.e., at a point in time, but not longitudinal in nature) and will provide associational, in contrast to causal, relationships. Nevertheless, the conceptual similarity of both approaches needs to be recognized and investigated in order to consider the possible integration of these methods. It should be noted that the principal references on the AET method are in two reports (Tetra Tech 1985, 1986) which do not have wide circulation and therefore, severely restrict the capacity of the broader scientific community to review this approach.

The SLC method to derive the SQC has considerable similarity to the ecosystem methodologies recommended by Kooijman (1987), Van Straalen and Denneman (1989), and Okkerman et al. (1991) in that the SLC and the collective Netherland ecosystem approaches are designed to estimate the distributional response of sample populations. However, the actual way in which this is achieved is quite different for each approach. The SLC approach offers to protect 95% of the observed (N = 20) species, while Van Straalen and Denneman (1989) attempt to estimate the NOEC for 95% of the

Table 7.4. Sediment Quality Criteria for Lead, Combustion Polyaromatic
Hydrocarbons (CPAHs), and Total Polychlorinated Biphenyls

	Criteria		
Criteria Descriptions	Pb ($\mu g/g$)	CPAHs ($\mu g/g$)	Total PCBs ($\mu g/g$)
No or minimal biological effects	≤ 50	≤ 3.8	≤ 0.1
Major biological effects	≥ 130	≥ 6.8	≥ 0.8
Area of uncertainty	$>50<130$	$>3.8<6.8$	$>0.1<0.8$

Source: P. M. Chapman, 1986.

community species based on a logistic distribution of observed species responses.

It should be noted that the EP approach is limited to estimating
effects-based criteria for the protection of benthic organisms. In a
recent major critical assessment of the EP methodology, DiTorro
et al. (1991) asserted the use of the EP approach to derive sediment
criteria for the protection of human health, wildlife, and marketa-
bility of fish and shellfish necessitates that the equilibrium assumption also include the water column and water column organisms.
He stated further that this was an "untenable assumption" at
present. This conclusion was based on observations that water
column concentrations can be much lower than pore water concentrations if sufficient dilution flow is occurring. In addition,
upper-trophic level animals may have concentrations considerably
greater than equilibrium values (Connolly and Pedersen, 1988).
These observations could preclude the derivation of SQCs from
WQCs, as was performed for certain interim criteria (Cowan and
DiTorro, 1988). Despite such restrictions, DiTorro et al. (1991) suggest that organism lipid-to-sediment organic carbon ratios may be
a viable approach for estimating the concentration of contaminants
in benthic species when the assumption of equilibrium can be justified. At present, they argue that a site-specific investigation involving food chain modeling is the only defensible approach to assess
the effect of contaminated sediments on the body burdens of upper-
trophic level organisms (see Connolly, 1991).

The bottom line, then, is which approach to derive SQC is
presently the most preferred and which methodology is likely to
take precedence in the future. While this may appear to be the correct question to ask, it's like saying who is the most important player

on a *team*. The various methods have their respective strengths and limitations, including scientific, range of applicability, technical implementation, timing, and cost. The initial testing, such as performed by Tetra Tech (1986), has helped enormously to define where the methods need to be strengthened and/or changed. Given the inherent problem with establishing causality with the various SQC, it is likely that most ecological risk assessors will demand that more than one approach be applied to a particular setting, especially taking into account features of complementarity and redundancy. This will depend on the site-specific nature of the sediments to be evaluated.

Use of SQC in MATC Derivation Process

SQC should be considered the equivalent of a species-specific, chemical-specific MATC. This is because its various methodologies may be used to derive acceptable sediment contaminant concentrations based on one or more biologically-based, species-specific endpoints of concern. Thus, a SQC value could be available for direct incorporation into an ecosystem MATC methodology, as presented in the following chapter. However, the SQC derivation can also offer a value relating to a different level of biological organization than the species. For example, the SQC could be based on changes in community structure. In such cases the risk assessor must be clear that one compares apples with apples and not apples with oranges. More specifically, SQC based on community change criteria should not be used with species-specific data for chemical-specific ecosystem MATC derivation.

8

Chemical-Specific Ecosystem MATC

A. OVERVIEW

The development of water quality and soil/sediment criteria in the context of ecological risk assessment has been the objective of numerous groups and organizations (Barnthouse et al. 1990; Kooijman 1985, 1987; Van Straalen and Denneman 1989; Okkerman et al. 1991; Fordham and Reagan 1991; EPA 1984, 1986; Van Der Gaag et al. 1991). These above methodologies represent major contributions to various aspects of the ecological risk assessment process. The contributions of each methodology must be seen within the context of its defined objectives. For example, Barnthouse et al. (1990) set forth procedures to derive species-specific MATCs. Okkerman et al. (1991), Kooijman (1987), and Van Straalen and Denneman (1989) attempted to formulate a basis for ecosystem risk assessment, but limited their objectives to essentially soil invertebrates under simple exposure scenarios. They thereby avoided complex food web issues of bioaccumulation and biomagnification, as well as the concept of endangered species. More recent approaches by Fordham and Reagan (1991) have incorporated complex food web modeling and MATissueC values to derive environmental criteria. However, the approach of Fordham and Reagan (1991) omitted consideration of the entire ecosystem, since impacts on soil bacteria, fungi, and other soil microorganisms were not considered.

The current scheme is designed to build upon the previously cited contributions. The purpose is to establish a sufficiently broad perspective that enables linkage of the salient features of the above efforts to derive a functional and defensible broadly based methodology that yields chemical-specific ecosystem MATCs.

B. SPECIFIC APPROACHES

1. From the Netherlands

The approach of Van Straalen and Denneman (1989) was initially developed for the assessment of the effects of compounds on terrestrial environments based on the work of Kooijman (1987). Their approach provides an estimation of a hazardous concentration (HC) affecting p% of the species in a community. The HCp is the concentration at which a certain percentage of the species in the community may be exposed to an unacceptable level. This method differs from the earlier Kooijman (1987) approach, which is designed to provide an estimate of a hazardous concentration for the most sensitive species in a community. The Kooijman (1987) approach leads to a much lower acceptable concentration.

The Kooijman/Van Straalen-Denneman approaches assume:

- the distribution of NOEC values within a large community can be described by a log logistic function,
- that species have been tested and that they represent a random sample from the community.

The procedure of Van Straalen and Denneman (1989) restricted itself to performing a "limited" ecosystem risk assessment for soil invertebrates (earthworms, springtails, mites, woodlice, etc.) exposed to cadmium contaminated soil. No specific consideration was given to the concerns of soil bacteria and fungi, as well as to terrestrial vertebrates. Also, the issue of persistent organic chemicals, bioaccumulation, and biomagnification were not addressed.

An important feature of the Van Straalen and Denneman (1989) method was its normalization of the soil for organic matter and clay content via a regression model for each contaminant. In the case of cadmium (Cd), the following relationship was used:

$$Rcd = 0.4 + 0.007 (L + 3H)$$

where
Rcd = Cd—reference values in $\mu g/g$
L = lutum content (particles $<2\mu m$), in % (w/w)
H = organic matter content in % (w/w)

This formulation is based on the methodology of partition equilibrium that is used by the U.S. EPA to establish sediment criteria for nonionic organic contaminants (DiTorro et al., 1991). Thus, by the formula, soils with high clay and organic matter content will have a higher acceptable level of cadmium than soils lower in these characteristics.

In cases of Kooijman (1987) and Van Straalen and Denneman (1989), the random selection of representative species is a critical assumption. This, however, is unlikely to be satisfied since it requires prior identification of all species/genera, selection of the testable organisms, and then conducting the actual assays. Bias in the selection of test species could be crucial since it is known that species can vary over 1000-fold in response to toxicants (Blanck, 1984).

In the absence of knowing how many species exist within a given community, Kooijman (1987) cited ecological literature indicating that the number of species is a function of surface area of the habitat (i.e., the species-area curve). He indicates that it can be predicted by a mathematical formula with the form of the function as follows:

$$N(a) = ba^r$$

where a is the surface area, b is a constant related to the number of species in a unit of surface area (of dimension length-2r), and r is another (dimensionless) constant.

In a direct comparison between the methodology of Van Straalen and Denneman (1989) and Kooijman (1987), using the same seven species, the estimated acceptable environmental standard (i.e., ecosystem) was 10,000-fold lower with the Kooijman approach! The major issue of conflict between these two extrapolation approaches is that Van Straalen and Denneman (1989) believe that 95% protection is a reasonable requirement while Kooijman (1987) argues for the need to protect even the most sensitive species within the context of their logistic distribution. Okkerman et al. (1991) believe that the approach of Kooijman (1987) yields excessively low values. The principal point here that should not be missed is that both approaches are technically similar, while their application is affected by different protection philosophies. That is, the Kooijman (1987) approach is to provide protection to the most sensitive modeled species. While Van Straalen and Denneman (1989) believe this to be a laudable goal, they do not feel it is practical after taking expense and natural background into account.

2. Two EPA Approaches

Two chemical-specific, MATC-like approaches have been offered by U.S. EPA scientists, including a 1984 approach (Table 8.1) (i.e., EPA Approach #1) and the report of Stephan et al. (1985) (i.e., EPA Approach #2), which provides water quality criteria guideline methods. The Stephan et al. (1985) approach provides a threshold estimate of unacceptable effects for a community. It assumes a 95% protection of genera and is estimated from the lowest of three chronic values (i.e., the final chronic value (FCV) for animals, the final plant value (FPV) and the final residue value (FRV). The FCV is determined via chronic NOEC values from at least eight animal families (i.e., family Salmonidae in the class Osteichtyes, a second family in Osteichtyes, a third family in the phylum Chordata, a planktonic crustacean, a benthic crustacean, an insect, a family in a phylum other than Arthropoda or Chordata not yet used).

Based on the available chronic NOEC values, a species mean chronic value is calculated for each species. The geometric mean of all species mean chronic values together are employed to estimate the mean chronic value per family. The four lowest genera mean chronic values per family are then used to estimate the FCV via the use of a triangular distribution model.

In 1984 the EPA proposed (i.e., EPA Approach #1) the use of an "assessment factor" in order to derive a type of chemical specific ecosystem MATC (i.e., concern level). When acute data exist for 5 invertebrate and fish species, the EPA recommends using the most sensitive species and an assessment factor of 100 (Table 8.1).

3. Critique of the Methodologies

Okkerman et al. (1991) have critiqued the respective approaches and concluded that the methodologies of the EPA (1984) and Stephan et al. (1985) are less reliable than that of Van Straalen and Denneman (1989). They argue that the EPA (1984) approach has an insufficient scientific basis, while the Stephan et al. (1985) approach is markedly limited by its assumption of a triangular distribution. The use of a triangular distribution assumption was

Table 8.1.   Comparison of Extrapolation Procedure Assessment Factors Applied
to Derive Concern Levels (EPA, 1984)

Available Information	Assessment Factor
One acute LC_{50} or QSAR estimate	1000
Lowest of 5 acute LC_{50}s for invertebrates and fish	100
Lowest chronic NOEC value of the most sensitive species[a]	10
Field effect concern levels	1

[a]To be valid, a NOEC must be preceded by acute toxicity test on several species;
the most sensitive species should then be used in the chronic test(s).

criticized by Okkerman et al. (1991) since it implies that there is
a threshold value below which effects will not occur. A second limitation is that only the four most sensitive species are employed.

4. Factors Affecting Model Estimates: Number of Species and Interspecies Variation

The principal limitations of the Van Straalen and Denneman
(1989) method involve how representative the selected species are
for the ecosystem and the number of species to be studied. For example, Okkerman et al. (1991) indicated that the maximum UF/mini-
mum UF ratio for 8 compounds ranged from 17,000 when tests on
three taxonomic groups were used, to 12 and 1.7 for the same 8
agents when five and seven taxonomic groups, respectively, were
employed. The key feature with sample (i.e., species) number is
*based on a formula that is strongly influenced by the size of the standard
deviation and a derived constant (dm) that integrates the number of species with the percent of species deemed necessary to protect. Based on this
type of assessment, it is generally concluded that as the number of species
tested increases, the uncertainty, and therefore the application factor size,
decreases.*

There are several important deviations from the above generali-
zation in the data presented by Okkerman et al. (1991), and these
will be emphasized in the following example. Of the eight agents
that were tested, two (i.e., TPBS and PCP) had markedly higher
HC_5 values with three species (N = 3) than with 5 or 7 species.
This difference was 107-fold for PCP when comparing N = 3 to
N = 5. The explanation can be accounted for on the basis of the
reduced variability in the data provided on the three species versus the five. This points out how vital the representativeness of

the species selected are. The large apparent disparity of the 17,000 to 1.7-fold maximum to minimum ratios as noted above is also an artifact of this process. The two examples chosen to derive the maximum UF/minimum UF ratios for N = 3 and N = 7 use dimethoate and PCP. In the case of N = 3 they represented extreme cases where the variation among the species in HC_5 values was ~3200-fold for dimethoate and 3-fold for PCP. In the latter case the standard deviation was considerably less with N = 3 than N = 7. These factors markedly affected the very low value for dimethoate (0.00083 mg/L) and the higher value for PCB (14 mg/L), and thus the 17,000 ratio value.

A better comparison would be to consider the magnitude of the difference among the HC_5 values for the N = 3 and N = 7 groupings. Of the six compounds where the N = 7 yielded lower values, the differences ranged from a low of 2.3-fold to a high of 433.7-fold, with a mean value of 85-fold and a median value of 21.1-fold. Of the two cases where the N = 3 comparisons yielded higher HC_5 values, the differences were 5-(TPBS) and 107.6-(PCP) fold.

This analysis reveals that the magnitude of variability in the responses of species tested is critical to the predictive outcomes (i.e., HC_5) and that the number of species, while important, is not the dominant feature in the estimation of the application factor.

5. Intermodel Comparisons

While Okkerman et al. (1991) argue that the procedure of Van Straalen and Denneman (1989) is the most theoretically defensible, it is important to point out that a comparison of the calculated ECL values based on EPA (1984) and those of Van Straalen and Denneman (1989) had a range of variation from as low as 1.06- to 2.84-fold when N = 7. When the N equaled 3, the range in ECL and HC_5 was from 2.6- to 1020-fold, while when N = 5 the range was from 1.3- to 8.3-fold. Thus, the difference between the EPA (1984) "safety factor" method and the Van Straalen and Denneman (1991) method when the number of species was 5 and 7 was less than a factor of 10 for each case. The variation was much greater when the N = 3, as noted above. Therefore, the two approaches were essentially equivalent, as judged by similarity of values and approximate proportionality of which approach offered more conservative (i.e., lower) estimations, when the estimated values were based on either 5 or 7 species.

Given the above discussion, a variety of options exist to estimate acceptable ecosystem levels of contaminants. While each approach has unique strengths and weaknesses, none have been extensively studied nor validated. This lack of conclusiveness with respect to which method is "best" led Okkerman et al. (1991) to select the Van Straalen and Denneman (1989) method on theoretical grounds. While this is a defensible recommendation, the above analysis indicates that it is subject to difficulties in satisfying its own assumptions, which can have a profound influence in the final estimated value. Of considerable interest is that in a limited comparison with 5 or 7 species involving eight agents, the EPA UF method and the Van Straalen and Denneman (1989) method provided essentially equivalent estimates. Whether this striking similarity is a coincidence or a generalizable pattern is not known. Nonetheless, the Van Straalen and Denneman (1989) HC_5 value method represents a theoretically sound approach for the derivation of chemical-specific ecosystem MATCs, and is recommended here.

The major difference in these approaches is that of modeling the responses of representative site-specific species or genera to estimate the responses of a given number of species (Van Straalen and Denneman 1989; Kooijman 1987, Stephan et al. 1985), versus essentially selecting a fixed (EPA 1984) or a derivable value (Slooff et al. 1986; Blanck 1984).

Okkerman et al. (1991) argue correctly that the selection should be based on which is more toxicologically plausible. It appears that the Kooijman (1987), Van Straalen and Denneman (1989), and Stephan et al. (1985) methods offer important advantages in that they attempt to model responses of species/genera that are site-specific.

The clear disparity in the estimates of acceptable concentrations based on current methods presents a problem for regulatory and health agencies. Selection of a site-specific model would appear to possess the strongest ecotoxicological relevance. However, regardless of the ecologically relevant model used, severe limitations exist due to randomization issues, the number of species tested, the arbitrariness of the percent of species to be protected, and lack of model validation.

C. ROLE OF FOOD WEB MODELING

1. General Considerations

The approaches of Okkermann et al. (1991) and Van Straalen and Denneman (1989) do not directly address the issue of bioaccumulation and biomagnification across trophic levels via a food web analysis (Van Der Gaag et al. 1991). They have restricted their assessment to toxicities which are related to partition equilibrium processes between the soil/sediment and the animal species. The only references to possible trophic level movement of persistent agents was made by Kooijman (1987). He indicated that generic safety factors (SFs) were used when dealing with issues of trophic level transfer. More specifically, a value of 1000 was used as a SF for nondegradable compounds with a high octanol-water partition coefficient, while a 100-fold SF was used for readily degradable agents. ESE (1989) also recommended a qualitatively similar SF approach for dealing with persistent chemicals when food web modeling was not used. In the case of ESE (1989), a factor of 10 for agents with a high BCF (≥ 1000) and a factor of 5 for chemicals with a BCF of ≥ 300 to < 1000 were recommended.

ESE (1989) provided no estimation of the number of species in an ecosystem nor the capacity of the studied species variability to estimate the ecosystem species variability. The goal of the ESE (1989) approach is to protect only those species actually seen with a strong assumption that the greatest focus be given to so-called "sink species."

It appears from the discussion of Kooijman (1987) that the SF for chemicals based on persistence in the environment was divided into the mean LC_{50} of the test species to arrive at an acceptable environmental (i.e., soil) concentration. Kooijman (1987) compared his method based on a HCp of 0.1 and on $N = 1000$ species for eight agents. He found that the logistic-derived AF ranged from a low of 60-fold to 7.10^7-fold (Table 8.2). This wide range in AFs led him to conclude that traditional UF approaches "threaten our environment for most elements."

The two approaches compared by Kooijman (1987) offer radically different methods to derive acceptable environmental concentrations. One approach estimates an acceptable environmental concentration by offering a generic UF for chemicals with a high likelihood for persistence and exposure. The basis for the use of

Table 8.2.   Mean $\overline{X}_m$ and SD S_m of m Chronic ln EC_{50} Values Given in Adema et al. 1981. (Based on these values, hazardous concentrations for sensitive species have been calculated for $\delta = 0.1$ and $n = 1000$.)

Compound	$\overline{X}_m$ (ln mgl^{-1})	S_m	m	HCS (μgl^{-1})	Application Factor T
2,4-Dichloroaniline	1.13	1.33	10	0.4	7800
Tetrapropylenebenzene sulphonate	3.45	0.87	10	90.0	350
Tricresyl phosphate	−0.49	1.59	10	0.014	45000
2,6-Dimethylquinoline	2.59	0.60	10	222	60
2,4-Dinitrotoluene	1.12	1.41	10	0.23	13300
Diisopropylamine	4.60	1.40	10	7.96	12500
Potassium dichromate	0.88	2.68	10	3.10^{-5}	7.10^7

Source: Kooijman, 1987.

the factor of 1000 was not presented. The Kooijman (1987) approach, in contrast, is based on actual variation in response to the toxin by test species at assumed steady-state conditions. The variability of community-based species (ecosystem) is then estimated using the logistic model. The basic Kooijman (1987) approach, as modified by Van Straalen and Denneman (1989) to adjust for partitioning factors, represents the conceptual synthesis of the linkage of the toxicological perspective with environment fate within the context of deriving environmental criteria. However, it needs further linkage with complex food web modeling to deal more comprehensively with the environment.

2. The "Sink" Species Method for Ecological Risk Assessment

The principal methodology for ecological risk assessment at the RMA is based on the use of sink species via food chain modeling. This assumes that exposure levels protective of sink species will also protect all other species, including those species inherently more sensitive than the sink species at a similar exposure (mg/kg) (Fordham and Reagan, 1991). This is believed to be justified on the basis that the enhanced exposure in the sink species will more than compensate for any differential susceptibility to the agents of concern.

The first question to ask is what type of interspecies variation in susceptibility can one expect if the intent is to protect the biota of the ecosystem. Differential susceptibility at the ecosystem level has been inferentially addressed by Slooff et al. (1986) in their comparison of 35 species over 11 taxa. Their comparison involved a large series of binary comparisons which estimated interspecies differences in susceptibility. The most direct attempts to estimate toxicity at the ecosystem level have been by Kooijman (1985, 1987), Van Straalen and Denneman (1989), and Okkerman et al. (1991) via the use of the logistic model. The use of population modeling offers substantial advantages since it addresses population distributional responses, rewards studies in which the database is more substantial, and can select the percent of the population that is desired to be protected. Kooijman (1987) conducted a model-driven exercise designed to derive ecosystem exposure criteria that would protect the most sensitive species based on the observed distributional responses of 1000 species. In this study a wide range of application factors (i.e., 7.1×10^7) were calculated. The range of estimated AFs was dominated by the variation in species response depending on the agent. Thus, for some agents, such as 2,6-dimethylquinoline, little interspecies variation was predicted, but for other agents, AFs of well over 10,000 were estimated (Table 8.2).

The question of differential susceptibility must also be seen within the context of species variation in exposure. This has direct relevance to the sink species concept. Within this context, predicted tissue contaminant concentrations (Table H3, page 3 of 3) by HLA (1991-part 4) of aquatic species at the RMA from the calibrated BMF values revealed the range of tissue concentrations for the following agents: 34.1-fold (algae/invertebrates to bald eagle) for dieldrin; 544-fold (algae to the bald eagle) and 109.1-fold (invertebrates to the bald eagle) for DDT; 474-fold (algae to bald eagle) and 156-fold (invertebrate to the bald eagle) for DDE. In addition, the bald eagle was estimated to bioaccumulate 39.4-fold more DDE, 29.6-fold more DDT, and 5.6-fold more dieldrin than the small fish; 22.6-fold more DDE, and DDT and 6.0-fold more dieldrin than the mallard duck; 19.1-fold more DDE, 13.8-fold more DDT, and 3.3-fold more dieldrin than the large fish. With respect to the terrestrial food web of the RMA (Table H2, page 2 of 3) (HLA 1991-part 4), the variation in predicted tissue concentrations of prairie dogs and the great horned owl is 5.3-, 3.3-, 1.95-, and 31.18-fold, for DDE, DDT, dieldrin and endrin, respectively.

While it is not possible to directly equate exposure levels (i.e., environmental concentration) with tissue concentration, it appears from this limited comparison that the premise of using sink species to compensate for interspecies variation in response to toxic substances may be seriously flawed. The basic flaw of the adequacy of sink species hypothesis lies in the fact that interspecies differences in susceptibility to toxic agents are frequently greater than 100-fold and occasionally exceed 1,000-fold (Slooff et al. 1986). At the same time, the degree of differential bioaccumulation in the above food web may be far below 50-fold. Thus, the assumption of Fordham and Reagan (1991) that the estimation of environmental median criteria via the use of sink species in the food web analysis method is inherently conservative and assures the protection of more sensitive species in the ecosystem is not convincing and likely to be frequently violated.

In the off-post (RMA) species identification, the HLA (1991, page 4) identified 22 reptile, six amphibian, and 44 bird species on whom little toxicological data exist. Tabulation of other species, including microorganisms, was not made. However, the total number of species in the RMA ecosystem is likely to be substantially greater than the 70 species of reptiles, amphibians, and birds listed. What the differential susceptibility of these species would be to toxic agents at RMA is unknown. The principal question, therefore, is whether the sink species pathway analysis method adequately compensates for the magnitude of variation in response to toxic substances seen within the ecosystem. The best answer at present is that it is likely to be only partially effective in deriving acceptable safe environmental criteria. The data and analyses presented provide little confidence in this approach per se. It is principally based on a hope that this procedure will work, but no validation exists that it is appropriate for a particular ecosystem. The sink species approach has a role in the spectrum of ecological risk assessment methodologies, but one that is not able to stand alone. It must be used as a complement to other approaches.

D. KEYSTONE SPECIES AND ECOSYSTEM MATC DERIVATION

The role of keystone species in ecological risk assessment remains elusive and in need of serious consideration. The answer to this role-defining question for keystone species will be a function of how and when ecological risk assessments are employed. The most obvious use of ecological risk assessment is at a site that is already contaminated, such as a Superfund site in the United States. Such sites have most likely been contaminated for numerous years with a complex set of chemical contaminants. A second case for ecological risk assessment could be as part of the already well-known process deemed an environmental impact statement. In the first instance the ecological risk assessment is conducted "after the fact" in an effort to define current and predict future adverse effects at the site in question. It is also concerned with providing decision-makers with accurate predictions of the ecological impacts of contaminants at various target concentrations that could be achieved via the application of different remediation strategies. In the case of a contaminated site it may not be possible to know what the natural community makeup was. What the assessors are confronted with is an assessment of what comprises the community at present. If species were deleted, if former minor species are now more dominant, or if new pollutant resistant species have become part of the community, it will not be known with a certainty. The principal focus of ecological risk assessment at contaminated sites would generally be expected to start at this point.

The questions that an ecological risk assessor may ask will usually be related to three principal toxicological endpoints (i.e., effects on growth, maintenance, and reproduction). As relevant as these questions may be within a laboratory setting, they do not speak directly to the issue of performance within food webs. This issue has not been adequately addressed nor even formally raised to the level of serious debate. Regardless of the need for toxicological endpoint-ecosystem data linkage, the basic reality is that regulatory agencies are facing a situation where a decision must be made as to whether the ecosystem has been adversely impacted and, if so, to what extent the contamination must be removed to prevent adverse effects from continuing. Since the ecological evaluation is an after-the-fact situation, no quantitative impact on the natural ecosystem will be available. It may be possible to assess the continuing exposure on the present community structure. However, it will not be known with confidence whether the surviving community represents the natural community or a modified, more pollutant-resistant community due to contaminant selection processes.

Since this scenario may typify many ecological risk assessments, the evaluation may underestimate the degree of contaminant-induced alterations that may have occurred. The predominant focus in such assessment is on chemically-related direct cause-effect relationships. However, what may be overlooked are critical secondary cause-effect relationships that may be even more profound to the function and survival of the ecosystem. The subsequent section addresses the differential impact on community structure, function, and integrity, depending on which particular type of species is most sensitive to contamination and is eliminated.

In 1966, Paine put forth the concept of a "keystone" species. These are individual populations that play a central and determining role in community structure, integrity, and stability. Removal of keystone species may profoundly alter the community composition affecting species diversity and composition (Paine, 1966).

The initial research of Paine (1966) determined that local species diversity is directly related to the efficiency with which predators prevent the monopolization of the major environmental requisites (e.g., space, food) by one species. In his 1966 study, Paine reported that experimental removal of the predator (i.e., starfish Pisaster) led to local extinctions of certain invertebrates and algae. More specifically, following the removal of the predator, the mussel Mytilus eliminated most of the other species in the community via competition or loss of food supply. Under normal circumstances the starfish eats large quantities of barnacles and thereby enhances the capacity of other species to survive in the area by allowing the space to be open. Paine (1966) concluded that when the top predator is artificially removed, the systems converge toward simplicity. Local diversity, therefore, on intertidal rocky bottoms was directly associated with predation intensity. In this situation the starfish was seen as a keystone species.

While Paine (1966, 1980) focused on the role of predation and species diversity within the concept of keystone species, Pimm (1980) expanded this concept by assessing the impact of selective removal of various types of species such as predators, prey, and plants on the community. Table 8.3 provides 19 examples of species deletion studies caused by natural or artificial (experimental) factors. Each example, which had its food web explicitly described, consisted of studies of freshwater pond and rocky intertidal habitats. While examples from terrestrial systems were not included because the criteria dealing with complexity of analyzed food webs were not satisfied, the findings are believed to be generalizable to include terrestrial food webs. The examples provided in Table 8.3

Table 8.3. Effects of Removing Predators

Web Species Removed	How Removed	Predator's Prey	Effects of Predator Removal
1 *Enhydra* (Sea otter)	Man; local conditions—cannot feed in deep water	*Strongylocentrotus* (Sea-urchin), fish and molluscs	*Strongylocentrotus* increases dramatically
2 *Strongylocentrus polyacanthus*	*Enhydra*	*Laminaria* and a variety of other macro-algae	*Agrarum* and *Thalassiophylum* (Algae) that are resistant to the predator are out-competed by *Laminaria*. Increased competition eliminated various species of red algae and *Alaria* (Algae)
3 *Strongylocentrotus purpuratus*	*Pycnopodia* (Starfish), *Anthopleura* (Anemone), local conditions—strong wave action, experimentally	Drift and *Hedophyllum* (Algae)	With predators, only coralline algae survive. Predator removal allows *Hedophyllum*, and the species associated with it to survive
4 As 3	Experimentally	Not given, but presumably as 3	*Lithothamnion* (Coralline algae) survives in presence of predator, but is out-competed by *Hedophyllum* when predator is removed
5 *Strongylocentrotus frasiscanni*	Experimentally	*Laminaria* is only species listed, but predator can survive in its absence	*Lithothamnion* survives in the presence of the predator but is out-competed by *Laminaria*, and overgrown by diatoms, *Ulva* and *Halosaccion* (Red algae) when predators are removed

continued

Table 8.3. *Continued*

Web Species Removed	How Removed	Predator's Prey	Effects of Predator Removal
6 *Echinometra, Diadema, Heterocentrotus* (Sea urchins)	Local conditions: wave action	Algal lawn	With predators corals survive, without them they are out-competed by the algae
7 *Paracentrotus* (Sea urchin)	Experimentally, local conditions not fully understood	Algae	A variety of algal species and the animals dependent on them invade essentially bare areas following predator removal
8 *Stichaster* (Starfish)	Experimentally	Specialized: 76% of the prey are *Perna* (Mussel)	"Rapid domination . . . by *Perna* . . . at the expense of the other resident, space-requiring species"
9 *Acanthaster* (Starfish)	Various coral symbionts (*Trapezia, Alpheus*) prevent predator from feeding on coral	Prefers *Pocillopora* (Branching corals), but will take other non-branching species	*Pocillopora* corals persist only in absence of predator, otherwise they are grazed close, or to, extinction
10 *Pisaster* (Starfish)	Experimentally	Very generalized	Predator removal allows *Mytilus* (mussels) to remove most of the other species in the community through competition or loss of food supply
11 *Pisaster*	Experimentally	As 10	Predator removal did not lead to increase in *Mytilus*. Larval *Mytilus* were probably scarce in the plankton during this experiment

continued

Table 8.3. *Continued*

Web Species Removed	How Removed	Predator's Prey	Effects of Predator Removal
12 *Katharina* (Chiton)	Experimentally	*Hedophyllum* and other algae	Presence or absence does not affect the rate at which the algae recovers when it, too, has been removed
13 *Acmaea* (Limpet)	Experimentally	Coralline algae	One change within the 5 species of coralline algae; one species became more common at the expense of another. Relative abundance of coralline algae versus *Hedophyllum* unchanged.
14 *Patella* (Limpet)	Experimentally	*Algae*	Dramatic increase in the abundance of the algae
15 *Lepomis* (Fish)	Experimentally, local conditions: dense vegetation	Principally larger zooplankton but will take smaller species if larger ones are unavailable	A large species, *Ceriodaphnia* (Cladoceran), predominated after predator removal at levels from 44% to 97% depending on nutrient levels in the pond. Smaller species (e.g., *Bosmina*, also a Cladoceran) and other herbivores (Rotifers and Copepods) are eliminated through competition
16 Odonates, coleopterans and other planktonic predators	Experimentally	*Daphnia*	Largest species, *Daphnia* (Cladoceran) increases, smaller species decrease; data are described as "weak" and one predator, *Chaoborus* (Dipteran) was not removed

continued

Table 8.3. *Continued*

Web Species Removed	How Removed	Predator's Prey	Effects of Predator Removal
17 *Anax* (Naid) (Benthic predator)	Experimentally	Mainly *Chironomus* (Dipteran), but also *Caenis* (Ephemeropteran)	*Chironomus* increases, *Caenis* decreases
18 *Chaoborus*	Experimentally	*Ceriodaphnia, Bosmina, Cyclops, Diaptomus* (last two are Cyclopoda); under strong predation pressure, *Daphnia*	At high densities *Chaoborus* feeds on rotifers and *Daphnia*, at low densities, smaller species (*Ceriodaphnia, Bosmina, Diaptomua*) can persist, but the rotifer predator, *Asplancha* (also a rotifer) is lost
19 *Lepomis* (Fish)	Experimentally	Four species of *Daphnia, Chaoborus, Cariodaphnia, Cyclops, Bosmina, Diaptomus*	With predator, smaller species (*Bosmina, Cyclops,* rotifers) predominate; the rotifers support *Asplancha.* Predator removal results in larger species (*Daphnia, Ceriodaphnia,* the predator, *Chaoborus*). Rotifers are not sufficiently abundant to support *Asplancha*

Source: Pemm, 1980.

are characterized by a large per capita effect of the predator on its prey, with only a small reverse effect. Based on these data Pimm (1980) concluded that real food webs are quite sensitive to the deletion of predators, since only four systems (i.e., numbers 11, 12, 13, and 16) of the 19 examples did not lose species following the removal or partial removal of a predator.

Modeling was also employed by Pimm (1980) to further assess the concept of species deletion (i.e., extinction) and its impact on a wide range of food webs. For example, he explored the question: "Within a web, which species can or cannot be removed (i.e., predator, prey, plants, etc.) without additional losses of species from the web?" This analysis involved 10 different food web types, each having 12 species with multiple predator-prey interactions. Based on this set of hypothetical food webs the following general conclusions were derived:

1. Removal of a species will cause most species losses from the species in the trophic levels beneath it when predators have a controlling effect on the equilibrium densities of their prey and those prey have a generalized choice of prey. Restating this in terms of competition: predator removal will have most effect when the prey are most extensively engaged in competition.
2. Removal of a species will have the least effect on the species in the trophic levels below it when either:
   a. predators exert a controlling influence on the equilibrium density of their prey, but those prey are specialized or not involved in competition, or
   b. the predators do not exert a controlling influence on the equilibrium densities of their prey.
3. Whether a predator is generalized or specialized will not affect whether its removal will cause further species losses.
4. Removal of a species will have the most marked effect on the species in the trophic levels above it when its predators are specialized on the species removed.

Based on the above assessment it is predicted that if keystone species are deleted as a result of enhanced susceptibility to contaminants, then this would have far-reaching impact on the community structure. As important as this concept is, there is no clear way to reconstruct this potential multiplicative impact. However, the loss of the keystone species has an effect far beyond itself. Consequently, the keystone species concept will be most effectively employed when ecological risk assessments are used in a truly predictive sense, such as in the case when emission standards from new point sources are being proposed.

The keystone species concept and the approach used in deriving chemical-specific ecosystem MATCs in the Netherlands seem difficult to reconcile. For example, the derivation of a chemical-specific ecosystem MATC by Van Straalen and Denneman (1989) and adopted by the Netherlands (Okkerman et al. 1991) assumes that up to 5% of the species are not intended to be protected with a NOEC. This was deemed as a practical necessity since protection of all species would likely yield such low acceptable values that it could not be financially achieved. Yet, the concept of automatically not building in protection for the lowest 5% of the species on the distribution has important ecological implications.

9

Deriving Multicontaminant
Ecosystem MATCs

While the previous chapters have focused on the derivation of chemical-specific, species-specific and/or ecosystem MATCs, the critical question is how to relate such efforts to the basic reality of waste sites with dozens of contaminants. The approach of Van Straalen and Denneman (1989) determines the chronic NOEL for the lower 5% of the species distribution. Their recommendations did not provide a means to account for the co-presence of multiple contaminants as would typically be the case. Their approach would derive chemical-specific ecosystem MATCs. Assume that there were 100 species in the ecosystem and 50 contaminants for which chemical-specific MATCs have been derived. If we assume further that susceptibility to each contaminant was randomly distributed among the species present, then considerably more than 5% of the species will be adversely affected as the number of contaminants increased.

The research of Blanck et al. (1984a) indicates that species susceptibility to toxic agents is highly variable and more consistent with the assumption of independence. If this is the case, then the application of the Van Straalen and Denneman (1989) method at a site where multiple contaminants exist would be predicted to permit a higher proportion of species being adversely affected.

If the goal of the ecosystem MATC for a particular site was not to allow 95% of the species present to be adversely affected by the contaminants present, then it would be necessary to estimate the collective species response to all the agents simultaneously. Thus, the greater the number of agents at the site, the more restrictive the chemical-specific ecosystem MATC would need to be. This could be accomplished with a Bonferoni-type correction procedure by which the value of 0.05 (i.e., level of protection) will be divided by the number of chemicals of concern present. For example, if 10 agents are present, each chemical-specific ecosystem MATC would now be calculated to provide protection to (.05/10 = .005) 99.5% of the population in order to ensure the overall ecosystem protection at the 0.05 level.

The practical implication of this approach reveals that in order to achieve the goal of ≤5% of the species being adversely affected, the individual chemical-specific ecosystem MATC would have to be considerably reduced. For example, changing from 95% to 99% to 99.9% protection could result in lowering the chemical-specific ecosystem MATC considerably. This is demonstrated in Example #7 in Appendix 2 where the UF was 29 at the 95% level, 197 at the 99%, and 2815 at the 99.9% level.

This generic approach does not address the issue of chemical-chemical interactions with respect to additivity, antagonism, and synergy. If data were available on the effects of multiple chemical exposures on specific species then this information could be incorporated into the scheme proposed in this chapter for deriving ecosystem MATCs for multiple toxicants. However, this would be performed on a case-by-case basis. Appendix 3 provides an assessment of how one may approach the assessment of mixture toxicology with ecological settings.

10

Final Thoughts

The preceding chapters have set forth a series of operational procedures along with theoretical underpinnings for the conduct of ecological risk assessments in a wide range of settings. The fact that a procedure exists and has a defensible toxicological foundation is gratifying, but it does not speak to the degree of confidence that exists in the predictions of the assessment. Although confidence in estimates of the risk assessment is a function of the nature of the risk assessment process, it is even more importantly dependent on the nature of the available data that can be used in the assessment. For example, if the risk assessor has only LC^{50} data on a single species available, any estimate of acceptable ecological risk is going to be highly uncertain, regardless of the extrapolation procedures adopted. If, on the other hand, chronic MATCs exist on multiple species that are representative of the site, then more confidence in a finally derived ecosystem value will exist regardless of the extrapolation procedure. The most critical feature in all risk assessments, therefore, is that of the quality, quantity, and relevance of the database. Too often there is extraordinary emphasis placed on extrapolation procedures and the derivation of default values in the absence of data, with less attention given to enhancing the quality of the database itself.

The principal orientation of both government and industry should be to encourage the creation of data from properly conceived, designed, and executed studies. This will result in mutual benefits to both entities, as well as the general public.

While it is essential to strongly advocate for the creation of appropriate data, the ecological risk assessor has to work with the data that are available, whether they are very limited or voluminous. The procedures recommended here are designed to accommodate the range of possibilities. While such flexibility is viewed here as a positive feature of the recommended procedures, it is likely that various organizations and individuals will object to how the methodology has dealt, for example, with the issue of uncertainties. More specifically, it would not be unexpected that acute toxicity data were treated with a 100 UF to approximate the chronic NOEL for the species on which the data were based. However, should this also be the case for other nontested species that are part of the ecosystem being studied?

Where differences of opinion reside may be seen in the following example. Current data suggest that about 90% of chemicals studied can have their chronic MATC estimated with an AF of 100-fold, while most of the remaining 10% of chemicals fall between the 100- and 1,000-fold AF. Thus, a value judgment would enter into the decision of whether to select 100 or 1,000 as the AF value. It should be emphasized that the distribution of the values previously noted appear to be at least in part a function of the type of agent, with metals and pesticides having larger AFs. If this were shown to be consistently true, then perhaps a case could be made for different-sized AFs according to chemical class. This position, while recently advocated in the literature (Van der Gaag et al. 1991), was not adopted in the present methodology. However, the earlier research of Kenaga (1978, 1982) provides support for this suggestion.

While the data of Kenaga (1978, 1982) related directly to the issue of high to low dose extrapolation, they do not address the issue of potential interspecies variation. Those advocating the use of the 100-fold AF as noted above, even when the species is not the one of interest, make the assumption that the species of interest and the species on which data exist are similarly susceptible. While this is not an unreasonable position, one has to assume that there is a 50% chance that the species of interest will be more sensitive than the species on which one has the data. The issue should not be whether the species of interest should be given some type of uncertainty consideration, but the magnitude of the factor. The experience of human risk assessment is that a 10-fold interspecies UF has helped to adequately assure that the human would not be adversely affected from chronic effects. Despite the long years of usage, whether this factor is appropriate is more a matter of belief than one of validation. It is predicated on the assumption that the animal model and human respond at least qualitatively similarly and that the only differences may be quantitative variation with similar organ systems affected. Given the general inadequacy of human clinical and epidemiological research with respect to pollutant-induced chronic endpoints, this assumption remains questionable. The size, therefore, of the interspecies UF for extrapolation to humans remains a somewhat conservative value since it is at least predicated on the belief that humans are more sensitive than the predictive model by one order of magnitude.

In the case of ecological toxicology, the direct comparative susceptibility of different species has been the object of extensive acute toxicity testing. The judgment emerging is that interspecies variation can be of considerable magnitude, with values exceeding 100-fold not uncommon even among relatively closely related species. The jury is still out on whether similar magnitudes of interspecies variation would exist for chronic endpoints as well.

The use of generic UFs is always intellectually dissatisfying because of its default basis. However, the present scheme attempted to improve this by applying a concept developed by Slooff et al. (1986) which allowed one to derive what may be called a tailored species UF. If the species of interest and the species on which the data exist for chemical X also had been tested in the past with many different agents, then a predictive relationship could be derived, including a 95% UF. This is a very attractive option for UF derivation because it offers a toxicologically-based value. However, this approach requires careful scrutiny on a case-specific basis prior to use. For example, the sample size is important, along with the types of agents used to establish the mathematical relationship. If the majority of the agents are metals and PAHs, and the COC is a chlorinated solvent, then the predictive model might not be very useful. It would be more relevant if the data were of agents in the same chemical class. Consequently, each time this approach is used it should be carefully assessed and justified. Nonetheless, its attractive features remain and it offers the potential for a significant advance over generic UFs.

The procedure for deriving the MATC has been one of some confusion in the field of ecological risk assessment. The term has been represented by a range (i.e., NOEC to LOEC), by the NOEC, and by the geometric mean of these two values. In human risk assessment procedures the comparable approach would be to select the equivalent of the highest NOEC. However, the procedure should estimate as closely as possible the threshold response and then employ this value as the MATC. The reason for using the threshold concept as the gold-standard here is that study designs can be highly variable and there is no guarantee that the highest NOEC will be close to the threshold, especially if the dosage spacing is wide (e.g., factors of 10). The use of the geometric mean is a procedure that can be used to estimate the true threshold. It is possible that this procedure could both over- and underestimate the true threshold.

The assumption of a log-normal distribution builds in some degree of conservatism. Whether one should use the highest NOEL or the geometric mean of the highest NOEL and the LOEL is not likely to be a major quantitative issue and should not be a troubling point. Selection of the highest NOEL assures the protection at the expense of achieving the goal of the best estimate of the threshold. Use of the geometric mean minimizes the magnitude of potential error and enhances the likelihood of estimating a value closer to the true threshold, although error may occur in estimating a value greater than the threshold.

This book has emphasized that ecological and human risk assessment share essential components, but principally differ in intent of the former to protect the population and the latter, to protect individuals. Both forms of risk assessment require exposure and hazard assessments. There are challenges, though, that are unique for the ecological domain. Human risk assessment focuses on one species and its distribution of exposure assumptions. It also uses a wide variety of models to estimate human responses with accompanying UFs. The ecological domain ideally will use a complex food web model that requires reliable procedures for calibration and validation for a particular site. The modeling procedure is extremely data-intensive and requires special technical skills with respect to characterizing the site chemically, biologically, and ecologically, which includes proper statistical sampling and analyses. It requires a recognition of the assumptions of the models and the uncertainties inherent in their use. The construction of a reliably predictable complex food web model is a time-consuming, highly expensive effort that still needs to be seen as a model of a dynamic (i.e., ever-changing) site.

Despite these challenges, the food web model is the link between the abiotic and biotic domains and represents a critical tool for the ecological risk assessor. Its use in site assessment has been emphasized here. The food web model acquires even greater significance when linked to toxicological and pharmacokinetic data that permit accurate predictions of tissue levels at dose rates for NOAELs and LOAELs. This information may then lead to straightforward procedures to determine acceptable environmental criteria.

Despite the apparent ease with which such calculations can be made, it is essential to avoid the easy trap of uncritically accepting one's estimates. Such assessments need to be critically reviewed for each assumption and their potential to quantitatively impact the final estimates. The Tetra Tech (1986) report on the confidence interval (CI) boundaries of the SQC estimated by the EP method for Puget Sound is illustrative of this point (see Chapter 7).

Equally important is the need to address the manner by which reliable and representative toxicological data can be obtained and used in ecosystem evaluations. The current schemes such as those from the Netherlands (Okkerman et al. 1991) and the U.S. EPA (Stephan et al. 1985) are recognized as inherently limited due to the issue of sample representativeness. While values obtained from such estimates are of great value since they represent the advanced deliberations of experts, they are not without considerable uncertainty, due principally to the formidable challenges confronting the ecological risk assessor. For example, Okkerman et al. (1991) require that the NOEC must be from species that are representative of the site. Obtaining data on truly representative species at each site of concern is probably an impossibility. Yet, if one of the basic assumptions of one's approach is never satisfied, any estimates based on that approach will be of a questionable accuracy. Such difficulties have not been resolved and will continue to be of concern to the ecological risk assessment process.

Despite this inherent difficulty and the numerous other limitations identified throughout the document, the process outlined is one that has both structure, organization, and form. It is the evolution of theoretical frameworks and experimental data from sediment and soil scientists, aquatic and terrestrial toxicology, and environmental modeling.

The human risk assessment process has taken center stage over the past two decades. It has been the object of detailed reviews by leading governmental experts within numerous countries and societies, as well as the National Academy of Sciences. However, it is expected that ecological risk assessment will find itself with ever increasing frequency under the same spotlight and will similarly profit from the deliberations of experts from many organizations. The present effort will hopefully contribute in a substantial way to these evolving and exciting deliberations that will work toward deriving a toxicologically-based international consensus on how to conduct ecological risk assessments.

References

Abbott, W. S. "A Method of Computing the Effectiveness of an Insecticide," *J. Econ. Entomol.* 18:265–267 (1925).

Abernethy, S. G., D. Mackay, and L. S. McCarty. " 'Volume Fraction' Correlation for Narcosis in Aquatic Organisms: The Key Role of Partitioning," *Environ. Toxicol. Chem.* 7:469–481 (1988).

Adams, W. J. "Toxicity and Bioconcentration of 2,3,7,8-TCDD to Fathead Minnows *(Pimephales promelas),*" *Chemosphere* 15:1503–1511 (1986).

Adema, D. M. M., and G. J. Vink. "A Comparative Study of the Toxicity of 1,1,2-Trichloroethane, Dieldrin, Pentachlorophenol and 3,4-Dichloroaniline for Marine and Fresh Water Organisms," *Chemosphere* 10:533–554 (1981).

American Society for Testing and Materials. *Estimating the Hazard of Chemical Substances to Aquatic Life.* (STP 657). J. Cairns, K. L. Dickson, and A. W. Maki, Eds. ASTM. Philadelphia, PA, 1978.

Barnthouse, L. W., G. W. Suter, II, and A. E. Rosen. "Risks of Toxic Contaminants to Exploited Fish Populations: Influence of Life History, Data Uncertainty and Exploitation Intensity," *Environ. Toxicol. Chem.,* 9:297–311 (1990).

Barnthouse L. W., G. W. Suter, II, A. E. Rosen, and J. J. Beauchamp. "Estimating Responses of Fish Populations to Toxic Contaminants," *Environ. Toxicol. Chem.,* 6:811–824 (1987).

Barnthouse, L. W., G. W. Suter, II, and A. E. Rosen. "Inferring Population-Level Significance from Individual-Level Effects: An Extrapolation from Fisheries Science to Ecotoxicology," in M. Lewis and G. W. Suter II, Eds., *Aquatic Toxicology and Environmental Fate,* 11th Vol., STP 1007. American Society for Testing and Materials, Philadelphia, PA, 1989, pp. 289–300.

Bartell, S. M., R. H. Gardner, and R. V. O'Neill. "Integrated Fates and Effects Model for Estimation of Risk in Aquatic Systems," in *Aquatic Toxicology and Hazard Assessment,* 10th Vol. ASTM STP 971. Philadelphia, PA, 1988, pp. 261–274.

Best, J. B. "Transphyletic Animal Similarities and Predictive Toxicology," in *Old and New Questions in Physics, Cosmology, Philosophy, and Theoretical Biology.* A. van der Merwe, Ed. (New York, NY: Plenum Press, 1983), pp. 549–591.

Blanck, H. "Species Dependent Variation Among Organisms in Their Sensitivity to Chemicals," *Ecol. Bull.*, 36:107–119 (1984).

Blanck, H., G. Wallin, and S.-A. Wangberg. "Species Dependent Variation in Algal Sensitivity to Chemical Compounds," *Ectotoxicol. Environ. Safety* 8:339–351 (1984a).

Blau, G. E., and W. B. Neely. "Mathematical Model Building with an Application to Determine the Distribution of Dursban Insecticide Added to a Simulated Ecosystem," *Adv. Ecol. Res.*, 9:133–163 (1975).

Borgmann, U., and D. M. Whittle. "Particle-Size–Conversion Efficiency and Contaminant Concentrations in Lake Ontario Biota," *Can. J. Fish Aquatic Sci.*, 40:328–336 (1983).

Breck, J. E., and S. M. Bartell. "Approaches to Modeling the Fate and Effects of Toxicants in Pelagic Systems," in *Toxic Contaminants and Ecosystem Health: A Great Lakes Focus*. M. S. Evans, Ed. (New York, NY: John Wiley & Sons, 1988), pp. 427–446.

Bringmann, G., and R. Kuhn. "Bergleich der toxischen Grenzkonzentrationen wassergefahrdender Stoffe gegen Bakterien, Algen und Protozoen im Zellvermehrungshemmtest," *Haustechnik-Bauphysik-Umwelttechnik*, 100:29–252 (1979).

Bringmann, G., and R. Kuhn. "Comparison of the Toxicity Thresholds of Water Pollutants to Bacteria, Algae, and Protozoa in the Cell Multiplication Inhibition Test," *Water Res.*, 14:231–241 (1980a).

Bringmann, G., and R. Kuhn. "Bestimmung der biologischen Schadwirkung wassergefahrdender Stoffe gegen Protozoen. II. Bakterienfressende Ciliaten-Z," *Wasser Abwasser Forsch.*, 1:26–31 (1980b).

Bringmann, G., and R. Kuhn. "Vergleich der Wirkung von Schadstoffen auf flagellate sowie ciliate bzw. auf holozoische bakterienfressende zowie saprozoische Protozoen," *Wasser Abwasser*, 122:308–313 (1981).

Brodie, B. B. Difficulties in extrapolating data on metabolism of drugs from animal to man. *Clin. Pharmacol. Ther.*, 3(3):374–380 (1962).

Burmaster, D. E., C. A. Menzie, J. S. Freshman, J. A. Burns, N. I. Maxwell, and S. R. Drew. "Assessment of Methods for Estimating Aquatic Hazards at Superfund Type Sites: A Cautionary Tale," *Environ. Chem. Tox.*, 10:827–842 (1991).

Burns, L. A., and G. L. Baughman. "Fate Modeling," in *Fundamentals of Aquatic Toxicology*. G. M. Rand and S. R. Petrocelli, Eds. Hemisphere Publishing Corp., 1985, pp. 558–584.

Calabrese, E. J. *Age and Susceptibility to Toxic Substances*. (New York, NY: John Wiley & Sons, 1986).

Calabrese, E. J. *Principles of Animal Extrapolation: Predicting Human Responses from Animal Studies*. (Chelsea, MI: Lewis Publishers, 1991).

Chapman, P. M. "Current Approaches to Developing Sediment Quality Criteria," *Environ. Toxic. and Chem.*, 8:589–599 (1989).

Chapman, P. M. "Sediment Quality Criteria from the Sediment Quality Triad: An Example," *Environ. Toxicol. Chem.*, 5:957–964 (1986).

Chapman, P. M., R. M. Dexter, and S. F. Cross. A Field Trial of the Sediment Quality Triad in San Francisco Bay, NOAA Technical Memo, NOS OMA 25. U.S. Department of Commerce, Rockville, MD, 1986.

Chapman, P. M., R. N. Dexter, and E. R. Long. "Synoptic Measures of Sediment Contamination, Toxicity and Infaunal Community Composition (the Sediment Quality Triad) in San Francisco Bay," *Mar. Ecol. Prog. Ser.*, 37:75–96 (1987).

Chiou, C. T., V. H. Freed, D. W. Schmedding, and R. L. Kohnert. "Partition Coefficient and Bioaccumulation of Selected Organic Chemicals," *Environ. Sci. Technol.*, 11:475–478 (1977).

Clark, T., K. Clark, S. Paterson, D. Mackay, and R. J. Norstrom. "Wildlife Monitoring Modeling, and Fugacity," *Environ. Sci. Technol.*, 22(2):120–127 (1988).

Connell, D. W., and G. Schuurmann. "Evaluation of Various Molecular Parameters as Predictors of Bioconcentration in Fish," *Ecotox. and Environ. Safety*, 15:324–335 (1988).

Connolly, J. P. "Application of a Food Chain Model to Polychlorinated Biphenyl Contamination of the Lobster and Winter Flounder Food Chains in New Bedford Harbor," *Environ. Sci. Technol.*, 25:760–770 (1991).

Connolly, J. P., and C. J. Pedersen. "A Thermodynamic-Based Evaluation of Organic Chemical Accumulation in Aquatic Organisms," *Environ. Sci. Tech.* 4, 22:99–103 (1988).

Cowan, C. E., and D. M. DiToro. Interim Sediment Criteria Valves for Nonpolar Hydrophobic Compounds. U.S. Environmental Protection Agency, Office of Water Regulations and Standards, Criteria and Standards Division, Washington, DC, 1988.

Crump, K. S. "A New Method for Determining Allowable Daily Intakes," *Fund. Appl. Toxicol.*, 4:84–87 (1984).

DiTorro, D. M., C. S. Zarba, D. J. Hansen, W. J. Berry, R. C. Swartz, C. E. Cowan, S. P. Pavlou, H. E. Allen, N. A. Thomas, and P. R. Paquin. "Technical Basis for Establishing Sediment Quality Criteria for Nonionic Organic Chemicals by Using Equilibrium Partitioning," *Environ. Toxicol. Chemistry*, 10(12):1541–1586 (1991).

Doherty, F. G. "Interspecies Correlations of Acute Aquatic Median Lethal Concentration for Four Standard Testing Species," *Environ. Sci. Technol.*, 17:661–665 (1983).

U.S. Environmental Protection Agency, "Estimating 'Concern Levels' for Concentrations of Chemical Substances in the Environment," in *Environmental Effects Branch Health and Environmental Review Division*, 1984.

U.S. Environmental Protection Agency, Interim Sediment Criteria Values for Non-Polar Hydropholac Organic Contaminants. SCD 17. Office of Water Regulations and Standards, Washington, DC, 1988.

U.S. Environmental Protection Agency, Superfund Public Health Evaluation Manual, EPA-540/1-86/060, 1986.

U.S. Environmental Protection Agency, Framework for Ecological Risk Assessment, EPA-630/R-92/001, Washington, DC, 1992.

U.S. Environmental Protection Agency, Peer Review Workshop Report on a Framework for Ecological Risk Assessment, EPA-625/3-91/022, Washington, DC, 1992a.

U.S. Environmental Protection Agency, Report on the Ecological Risk Assessment Guidelines Strategic Planning Workshop, EPA-630/R-92/002, Washington, DC, 1992b.

ESE. Biota Remedial Investigation. Final Report (Version 3.2), Vol. II. Environmental Sciences and Engineering, Inc., Rocky Mountain Arsenal, Commerce City, CO, 1989.

Fordham, C. L., and D. P. Reagan. "Pathways Analysis Method for Estimating Water and Sediment Criteria at Hazardous Waste Sites," *Environ. Toxic. & Chem.*, 10:949–960 (1991).

Fry, F. E. J. "The Physiology of Fishes," Vol. 1. M. E. Brown, Ed. (New York: Academic Press, 1957).

Gardner, H. "Biological Assessment of Contaminants in Surface and Groundwater," Presented at Conference on Drinking Water and Public Health, Amherst, MA, May 2, 1990.

Geyer, H. J., I. Scheunert, J. G. Filser, and F. Korte. Bioconcentration Potential (BCP) of 2,3,7,8-Tetrachlorodibenzo-p-dioxin (2,3,7,8-TCDD) in Terrestrial Organisms Including Humans," *Chemosphere*, 15(9–12):1495–1502 (1986).

HLA (Harding Lawson Associates Environmental Science and Engineering, Inc.) Revised Final Draft Report. Rocky Mountain Arsenal, Commerce City, Colorado, 1991.

Johnson, W. W., and M. T. Finley. Handbook of Acute Toxicity of Chemicals to Fish and Aquatic Invertebrates. U.S. Fish and Wildlife Service Resource Publication 137. U.S. Department of the Interior Washington, DC, 1980.

Katsuna, M., Ed. Minimata Disease. Kumamoto University, Japan, 1968. (Cited in Willes, 1977).

Kenaga, E. E. "Comparative Toxicity of 131,596 Chemicals to Plant Seeds," *Ecotoxicol. Environ. Safety*, 5:469–475 (1981).

Kenaga, E. E., and C. A. I. Goring. Relationship Between Water Solubility, Soil Sorption, Octanol-Water Partitioning, and Concentration of Chemicals in Biota," in *Aquatic Toxicology*, Third Symposium. ASTM-STP 707. J. G. Eaton, P. R. Parrish, and A. C. Hendricks, Eds. American Society for Testing and Materials, Philadelphia, PA, 1980, pp.78–115.

Kenaga, E. E., and R. J. Moolenaar. "Fish and Daphnia Toxicity as Surrogates for Aquatic Vascula Plants and Algae," *Environ. Sci. Technol.*, 13:1479–1480 (1979).

Kenaga, E. E. "Test Organisms and Methods Useful for the Early Assessment of Acute Toxicity of Chemicals," *Environ. Sci. Technol.*, 12:1322–1329 (1978).

Kenaga, E. E. "Aquatic Test Organisms and Methods Useful for Assessment of Chronic Toxicity of Chemicals," in K. L. Dickson, A. W. Maki, and J. Cairns, Jr., Eds., *Analyzing the Hazard Evaluation Process.* American Fisheries Society, Washington, DC, 1979, pp. 101–111.

Kenaga, E. E. "Predictability of Chronic Toxicity from Acute Toxicity of Chemicals in Fish and Aquatic Invertebrates," *Environ. Toxicol. Chem.,* 1:347–358 (1982).

Kester, J. E., M. I. Banton, and M. L. Stoltz. "Oral Bioavailability of Soil-Associated Aldrin/Dieldrin," *The Toxicologist,* 12(1):425 (abstract) (1992).

Kimerle, R. A., A. F. Werner, and W. J. Adams. "Aquatic Hazard Evaluation Principles Applied to the Development of Water Quality Criteria," in *Aquatic Toxicology and Hazard Assessment. STP 854. American Society for Testing and Materials,* Philadelphia, PA, 1983, pp. 538–547.

Kishino, T., and K. Kobayashi. "A Study on the Absorption Mechanism of Pentachlorophenol in Goldfish Relating to Its Distribution Between Solvents and Water," *Bull. Jpn. Soc. Sci. Fish,* 46:1165–1168 (1980).

Kobayashi, K., and T. Kishino. "Effect of pH on the Toxicity and Accumulation of Pentachlorophenol in Goldfish," *Bull. Jpn. Soc. Sci. Fish,* 46:167–170 (1980).

Konemann, H. "Structure-Activity Relationships and Additivity in Fish Toxicities of Environmental Pollutants," *Ecotoxicol. Environ. Safety,* 4:415–421 (1980).

Kooijman, S. A. L. M. "A Safety Factor for LC_{50} Values Allowing for Differences Among Species," *Wat. Res.,* 21(3) 269–276 (1987).

Kooijman, S. A. L. M. "Toxicity at Population Level," in *Multispecies Toxicity Testing* (edited by J. Cairns, Jr.) (New York, NY: Pergamon Press, 1985.)

Layton, D. W., B. J. Mallon, D. H. Rosenblatt, and M. J. Small. "Deriving Allowable Daily Intakes for Systemic Toxicants Lacking Chronic Toxicity Data," *Reg. Toxic. Pharm.,* 7:96–112 (1987).

LeBlanc, G. A. "Interspecies Relationships in Acute Toxicity of Chemicals to Aquatic Organisms," *Environ. Toxicol. Chem.,* 3:47–60 (1984).

Linden, E., B.-E. Bengtsson, O. Svanberg, and G. Sundstrom. "The Acute Toxicity of 78 Chemicals and Pesticide Formulations Against Two Brackish Water Organisms, the Bleak (*Alburnus alburnus*) and the Harpacticoid (*Nitocra spinipes*)," *Chemosphere,* 11/12:843–851 (1979).

Lloyd, R. "Effect of Dissolved Oxygen Concentration on the Toxicity of Several Poisons to Rainbow Trout (*Salmo gairdneri* Richardson)," *J. Exp. Biol.,* 38:447–455 (1961).

Long, E. R., and P. M. Chapman. "A Sediment Quality Triad: Measures of Sediment Contamination, Toxicity and Infaunal Community Composition in Puget Sound," *Mar. Pollut. Bull.,* 16:405–415 (1985).

Macek, K. J., S. R. Petrocelli, and B. H. Sleight III. "Considerations in Assessing the Potential for, and Significance of, Biomagnification of Chemical Residues in Aquatic Food Chains," in *Aquatic Toxicology,* L. L.

Marking and R. A. Kimerle, Eds. ASTM STP 667 (Philadelphia, PA: ASTM, 1979), pp. 251–268.

Macek, K. J., and B. H. Sleight. "Utility of Toxicity Tests with Embryos and Fry of Fish in Evaluating Hazards Associated with the Chronic Toxicity of Chemicals to Fishes," in *Aquatic Toxicology and Hazard Evaluation*. STP 634. American Society for Testing and Materials, Philadelphia, PA, 1977, pp. 137–146.

Mackay, D. *Multimedia Environmental Models: The Fugacity Approach* (Chelsea, MI: Lewis Publishers, 1991).

Mackay, D. "Correlation of Bioconcentration Factors," *Environ. Sci Technol.*, 16:274–278 (1982).

Mackay, D., and A. Hughes. "Three-Parameter Equation Describing the Uptake of Organic Compounds by Fish," *Environ. Sci. Tech.*, 18:439–444 (1984).

Mackay, D., and S. Paterson. "Fugacity Revisited: The Fugacity Approach to Environmental Transport," *Environ. Sci. Tech.*, 16:654A–660A (1982).

Mackay, D., P. V. Roberts, and J. A. Cherry. "Transport of Organic Contaminants in Groundwater," *Environ. Sci. Technol.*, 19:384–392 (1985).

Maki, A. W. "Correlations Between *Daphnia magna* and Fathead Minnow (*Pimephales promelas*) Chronic Toxicity Values for Several Classes of Test Substances," *J. Fish. Res. Board Can.*, 36:411–421 (1979).

Maki, A. W., and W. E. Bishop. "Chemical Safety Evaluation," in *Fundamentals of Aquatic Toxicology*, G. M. Rand and S. R. Petrocelli, Eds. (Bristol, PA: Hemisphere Publishing Corp., 1985), pp. 558–584.

Mayer, F. L., and R. H. Kramer. As cited in Neely, 1979. *Prog. Fish-Cult.*, 35:9 (1973).

McCarty, L. S. "The Relationship Between Aquatic Toxicity QSARS and Bioconcentration for Some Organic Chemicals," *Environ. Toxic. Chem.*, 5:1071–1080 (1986).

McCarty, L. S., P. V. Hodson, G. R. Graig, and K. L. E. Kaiser. "The Use of Quantitative Structure Activity Relationships to Predict the Acute and Chronic Toxicities of Organic Chemicals to Fish," *Environ. Toxic. Chem.*, 4:595–606 (1985).

McKim, J. M. "Evaluation of Tests with Early Life Stages of Fish for Predicting Long-Term Toxicity," *J. Fish. Res. Board Can.*, 34:1148–1154 (1977).

McNamara, B. P. "Concepts in Health Evaluation of Commercial and Industrial Chemicals," in *New Concepts in Safety Evaluation*, M. Mehlman, R. Shapiro, and H. Blumenthal, Eds. *Advances in Modern Toxicology*, Vol. 1. (1976) pp. 61–140.

Moriarty, F. "Bioaccumulation in Terrestrial Food Chains," in *Appraisal of Tests to Predict the Environmental Behavior of Chemicals*, P. Sheehan, F. Korte, W. Klein, and P. Bourdeco, Eds. (London: J. Wiley and Sons, Ltd., 1985), pp. 257–284.

Moriarty, F., and C. H. Walker. "Bioaccumulation in Food Chains—A Rational Approach," *Ecotox. Environ. Safety,* 13:208–215 (1987).

Mosberg, A. T., and A. W. Hayes. "Subchronic Toxicity Testing," in *Principles and Methods of Toxicology* (New York, NY: Raven Press, Ltd., 1989) p. 223.

Mount, D. I., and C. E. Stephan. "A Method for Establishing Acceptable Toxicant Limits for Fish-Malathion and the Butoxyethanol Ester of 2,4-D," *Trans. Am. Fish. Soc.,* 96:185–193 (1967).

Mount, D. I., and C. E. Stephan. "Chronic Toxicity of Copper to the Fathead Minnow (*Pimephales promelas*) in Soft Water," *J. Fish. Res. Board Can.,* 26:2449–2457 (1969).

Neely, W. B. "Estimating Rate Constants for the Uptake and Clearance of Chemicals by Fish," *Environ. Sci. and Technol.,* 13(12):1506–1510 (1979).

Neely, W. B., and G. E. Blau. *Pesticides in the Aquatic Environments,* M. A. Q. Khan, Ed. (New York: Plenum Publishers, 1977).

Neely, W. B., D. R. Branson, and G. E. Blau. "Partition Coefficient to Measure Bioconcentration Potential of Organic Chemicals in Fish," *Environ. Sci. Technol.,* 8(13):1113–1115 (1974).

Neff, J. M., B. W. Cornaby, R. M. Vaga, T. C. Gulbransen, J. A. Scanlon, and D. J. Bean. "An Evaluation of the Screening Level Concentration Approach to Derivation of Sediment Quality Criteria for Freshwater and Saltwater Ecosystems," in W. J. Adams, G. A. Chapman and W. G. Landis, Eds., *Aquatic Toxicology and Hazard Assessment: Tenth Volume.* STP 971. American Society for Testing and Materials, Philadelphia, PA, 1988, pp. 115–127.

Niederlehner, B. R., J. R. Pratt, A. L. Buikema, Jr., and J. Cairns, Jr. "Comparison of Estimates of Hazard Derived at Three Levels of Complexity," Community Toxicity Testing, ASTM STP 920, John Cairns, Jr., Ed., American Society for Testing and Materials,. Philadelphia, PA, 1986, pp. 30–48.

Norberg-King, T. J. "Predicting Chronic Toxicity with Fathead Minnows," *Environ. Toxicol. Chem.,* 8:1075–1089 (1989).

Norstrom, R. J., A. E. McKinnon, and A. S. W. DeFreitas. "A Bioenergetics-Based Model for Pollutant Accumulation by Fish. Simulations of PCB and Methylmercury Residue Levels in Ottawa River Yellow Perch (*Perca flavescens*)," *J. Fish. Res. Board Can.,* 33:248–267 (1976).

Okkerman, P. C., E. J. Plassche, W. Slooff, C. J. Van Leeuwen, and J. H. Canton. "Ecotoxicological Effects Assessment: A Comparison of Several Extrapolation Procedures," *Ecotoxicology and Environmental Safety,* 21:182–193 (1991).

Oliver, B. G., and A. J. Niimi. "Bioconcentration Factors of Some Halogenated Organics for Rainbow Trout: Limitations in Their Use for Prediction of Environmental Residues," *Environ. Sci. Technol.,* 19:842–849 (1985).

Paine, R. T. "Food Web Complexity and Species Diversity," *The American Naturalist* 100:65–93 (1966).

Paine, R. T. "Food Webs: Linkage, Interaction Strength and Community Ultrastructure," *J. Animal Ecology* 49:667–685 (1980).

Palmer, E. L. and H. S. Fowler. *Field of Natural History*, 2nd Ed. (NY: McGraw-Hill Book Co., 1975).

Patrick, R., J. Cairns, Jr., and A. Scheier. "The Relative Sensitivity of Diatoms, Snails, and Fish to Twenty Common Constituents of Industrial Wastes," *Progve. Fish Cult.*, July, 1968, pp. 137–140.

Pavlou, S. P. "The Use of the Equilibrium Partitioning Approach in Determining Safe Levels of Contaminants in Marine Sediments," in K. L. Dickson, A. W. Maki and W. A. Brungs, Eds., *Fate and Effects of Sediment-Bound Chemicals in Aquatic Systems*. (Toronto, Ontario: Pergamon Press, 1987) pp. 388–412.

Pearson, J. G., J. P. Glennon, J. J. Barkely, and J. W. Highfill. "An Approach to the Toxicological Evaluation of a Complex Industrial Wastewater," in L. L. Marking and R. A. Kimerle, Eds., *Aquatic Toxicology*. STP 667. American Society for Testing and Materials, Philadelphia, PA, 1979 pp. 284–301.

Pimm, S. L. "Food Web Design and the Effect of Species Deletion," *Oikos* 35:139–145 (1980).

Portier, C., A. Tritscher, M. Kohn, C. Sewall, G. Clark, L. Edler, D. Hoel, and G. Lucier (1993) "Ligand/Receptor Binding for 2,3,7,8-TCDD: Implications of Risk Assessment." *Fund. Appl. Toxicol.*, 20:48–56.

Rappaport, T. A., and S. J. Eisenreich. "Chromatographic Determination of Octanol-Water Partition Coefficients (Kows) for the 58 PCB Congeners," *Environ. Sci. Technol.*, 18:163–170 (1984).

Shea, D. "Developing National Sediment Quality Criteria," *Environ. Sci. Technol.*, 22:1256–1261 (1988).

Slooff, W. "Benthic Macroinvertebraes and Water Quality Assessment: Some Toxicological Considerations," *Aquat. Toxicol.*, 4:73–82 (1983).

Slooff, W., J. H. Canton, and J. L. M. Hermens. "Comparison of the Susceptibility of 22 Freshwater Species to 15 Chemical Compounds. I. (Sub)acute Toxicity Tests," *Aquat. Toxicol.*, 4:113–128 (1983).

Slooff, W., and J. H. Canton. "Comparison of the Susceptibility of 11 Freshwater Species to 8 Chemical Compounds. II. (Semi) Chronic Tests," *Aquat. Toxicol.*, 4:271–282 (1983).

Slooff, W., J. A. Van Oers, and D. DeZwart. "Margins of Uncertainty in Ecotoxicological Hazard Assessment," *Environ. Toxicol. Chem.*, 5:841–852 (1986).

Smith, I., and G. Craig. "Prediction of Organic Contaminant Aquatic Toxicity Utilizing Intraperitoneal Injection and Structure-Activity Relationships," *Can. Tech. Report Fish. Aquat. Sci.*, 1151:122–131 (1983).

Southworth, G. R., C. C. Keffer, and J. J. Beauchamp. "Potential and Realized Bioconcentration. A Comparison of Observed and Predicted Bioconcentration of Azaarene in the Fathead Minnow (*Pimephale promelas*)," *Environ. Sci. Technol.*, 14:1529 (1980).

Southworth, G. R., C. C. Keffer, and J. J. Beauchamp. "The Accumulation and Disposition of Benz(a)acridine in the Fathead Minnow, (*Pimephales promelas*)," *Arch. Environ. Contam. Toxicol.*, 10:561 (1981).

Spacie, A., and J. L. Hamelink. "Bioaccumulation," in *Fundamentals of Aquatic Toxicology*. G. M. Rand and S. R. Petrocelli, Eds. (Bristol, PA: Hemisphere Publishing Corporation, 1985), pp. 495-525.

Spehar, R., H. Nelson, M. Swanson, and R. Renoos. "Pentachlorophenol Toxicity to Amphipods and Fathead Minnows at Different Test pH Values," *Environ. Sci. Toxicol.*, 4:389-397 (1985).

Stephan, C. E., and J. W. Rogers. "Advantages of Using Regression Analysis to Calculate Results of Chronic Toxicity Tests," *Aquatic Toxicology and Hazard Assessment: Eighth Symposium*. ASTM STP 891. R. C. Bahner and D. J. Hansen, Eds., American Society for Testing and Materials, Philadelphia, PA, 1985, pp. 328-338.

Stephan, C. E., D. I. Mount, D. J. Hansen, J. H. Gentile, G. A. Chapman, and W. H. Brungs. "Guidelines for Deriving Numeric National Water Quality Criteria for the Protection of Aquatic Organisms and Their Uses," (NTIS PB85-227049). U.S. Environmental Protection Agency, Washington, DC, 1985.

Suter, G. W. II. "Seven-Day Tests and Chronic Tests," *Environ. Toxicol. Chemistry*, 9:1435-1436 (1990).

Suter, G. W. II, D. S. Vaughan, and R. H. Gardner. "Risk Assessment by Analysis of Extrapolation Error, A Demonstration for Effects of Pollutants on Fish," *Environ. Toxicol. Chem.*, 2:369-378 (1983a).

Suter, G. W., L. W. Barnthouse, J. E. Breck, R. H. Gardner, and R. V. O'Neill. "Extrapolation from the Laboratory to the Field: How Uncertain Are You?" in *Aquatic Toxicology and Hazard Assessment*. STP 854. American Society for Testing and Materials, Philadelphia, PA, 1983b, pp. 400-413.

Suter, G. W. II, L. W. Barnthouse, and R. V. O'Neill. "The Uses of Uncertainty in Environmental Risk Analysis" (1983c), in *Quantification of Risk, Reducing the Uncertainties, Proceedings of a Symposium*. Santa Barbara, CA, June 21-24, 1982.

Suter, G. W. II, A. E. Rosen, E. Linder, and D. F. Parkhurst. "Endpoints for Responses of Fish to Chronic Toxic Exposures," *Environ. Toxicol. Chem.*, 6:793-809 (1987).

Tetra Tech. "Commencement Bay Nearshore/Tidalflats Remedial Investigation," Report prepared for the Washington Department of Ecology and U.S. Environmental Protection Agency, Seattle, WA. EPA-910/9-85-134b. Tetra Tech, Bellevue, WA (1985).

Tetra Tech. "Commencement Bay Nearshore/Tidalflats Remedial Investigation," Report prepared for the Washington Department of Ecology and U.S. Environmental Protection Agency, Seattle, WA. EPA-910/9-85-134b. Tetra Tech, Bellevue, WA (1985).

Tetra Tech. "Development of Sediment Quality Values for Puget Sound," Report prepared for the U.S. Environmental Protection Agency, Washington Departments of Ecology and Natural Resources, and U.S. Army Corps of Engineers, Seattle District, Seattle, WA. Tetra Tech, Bellevue, WA (1986).

Tetra Tech. "Development of Sediment Quality Values for Puget Sound," Volumes 1 and 2. Tetra Tech, Inc. 11820 Northop Way, Suite 100, Bellevue, WA, 98005 (1986).

Thomann, R. V. "Equilibrium Model of Fate of Microcontaminants in Diverse Aquatic Food Chains," *Can. J. Fish. Aquat. Sci.*, 38:280–296 (1981).

Thomann, R. V. "Bioaccumulation Model of Organic Chemical Distribution in Aquatic Food Chains," *Environ. Sci. Technol.*, 23:699–707 (1989).

Thomann, R. V., J. P. Connolly, and T. F. Parkerton. "An Equilibrium Model of Organic Chemical Accumulation in Aquatic Food Webs with Sediment Interaction," *Environ. Chem. Tox.* 11:615–629 (1992).

Van Der Gaag, M. A., B. M. Stortelder, W. A. Bruggeman, and L. A. Van Der Koou. "Risks of Toxic Compounds in Aquatic Systems: Science and Practice," *Comp. Biochem. Physiol.* Vol. 100C No. 1/2, 1991, pp. 279–281.

Van der Schalie, W. H., and H. S. J. Gardner. "New Methods of Toxicity Assessment in Military Relevant Applications," U.S. Army Biomedical Research and Development Laboratory. Fort Detrick, MD (not dated).

Van Hoogen, G., and A. Opperhuizen, A. "Toxicokinetics of Chlorobenzenes of Fish," *Environ. Toxicol. Chem.*, 7:213–219 (1988).

Van Straalen, N. M., and C. A. J. Denneman. "Ecotoxicological Evaluation of Soil Quality Criteria," *Ecotoxicol. Environ. Saf.*, 18:241–251 (1989).

Veith, G., D. Call, and L. Brooke. "Structure-Toxicity Relationship for the Fathead Minnow, *Pimephales promelas:* Narcotic Industrial Chemicals," *Can. J. Fish. Aquat. Sci.*, 40:743–748 (1983).

Veith, G., D. DeFoe, and B. Bergstedt. "Measuring and Estimating the Bioconcentration Factor of Chemicals in Fish," *J. Fish Res. Board Can.*, 36:1040–1048 (1979).

Woltering, D. M. "The Growth Response in Fish Chronic and Early Life Stage Toxicity Tests: A Critical Review," *Aquat. Toxicol.*, 5:1–21 (1984).

Zeise, L., E. A. C. Crouch, and R. Wilson. "Safety Evaluation and Risk Assessment," *J. Amer. College Tox.*, 5(2):137–152 (1986).

Appendix 1

Resumé of Vertebrate Classification

[Source: *Principles of Zoology*, W. H. Johnson, L. E. Delanney, E. C. Williams, and T. A. Cole, Eds. (New York, NY: Holt, Rinehart and Winston, Inc., 1969), pp. 337–343.]

All vertebrates have the basic chordate characteristics and in addition they are distinguished from all other animals and from each other as follows (living groups only are given).

Subphylum Vertebrata (L. *vertebratus*, jointed). Possess a spinal cord surrounded by a bony or cartilaginous skeleton, the vertebral column. Brain encased in the protective cranium.

Class Agnatha (Gk. *a*, without + *gnathos* jaw). Cylindrical body, dorsal fin only, jaws absent, mouth suctorial with horny teeth, skin smooth and scaleless, gills located in pouches. Two orders, about 50 species. Lampreys (*Petromyzon*), slime eels (*Myxine*), and hagfishes (*Polistrema*).

Class Chondrichthyes (Gk. *chondros*, cartilage + *ichthys*, fish). Cartilaginous skeleton; leathery skin with placoid scales; jaws present; males with claspers; eggs large and with much yolk. Three orders, about 600 species. Sharks (*Squalus*), (*Charcharodon*); Rays (*Raja, Torpedo*); and Chimaeras or Ghost Fish (*Chimaera*).

Class Osteichthyes (Gk. *osteon*, bone + *icthys*, fish). Skeleton mostly bony; skin usually with imbedded dermal scales; paired lateral fins and median fins usually present; fins supported by fin rays; gills in single chambers on either side of the pharynx.

Subclass Choanichthyes (Gk. *choana*, funnel + *ichthys*, fish). Nostrils open into mouth cavity; paired fins with large lobe containing skeletal elements and muscles. Three orders, less than 10 species. Lungfish (*Lepidosiren*), Coelacanth (*Latimeria*).

Subclass Actinopterygii (Gk. *actis*, ray + *pteryx*, fin). Nostrils do not open into mouth cavity; fins without lobes and supported by dermal rays. Over 30 orders and more than 17,000 species. Some of the more common orders are:

Order Acipenseriformes (L. *acipenser*, sturgeon + *formis*, having the form of). Long snout; no teeth; cartilaginous skeleton. Sturgeon (*Acipenser*), paddlefish (*Polyodon*).

Order Clupeiformes (L. *clupea*, small river fish + *formis*, having the form of). Fins without spiny rays; pelvic fins posterior in abdominal region; duct of air bladder open to the pharynx. Tarpon (*Tarpon*), herring (*Clupea*), salmon (*Salmo* and *Onchorhynchus*), pike (*Esox*).

Order Cypriniformes (L. *cyprinus*, carp + *formis*, having the form of). Anterior vertebrae fused; chain of bones connects air bladder with inner ear; body may be covered with scales, or with bony plates, or be naked. Over 5000 species. Piranha (*Serrasalmus*), electric eel (*Electrophorus*), carp (*Cyprinus*), catfish (*Ictalurus*).

Order Anguilliformes (L. *anguilla*, eel + *formis*, having the form of). Body long and slender; dorsal, caudal, and anal fins continuous; scales vestigial or absent. Eels (*Anguilla*), morays (*Muraena*).

Order Gadiformes (L. *gadus*, cod + *formis*, having the form of). Fins soft-rayed; air bladder lacks duct; mostly marine and dwell on the bottom. Codfish (*Gadus*), haddock (*Melanogrammus*).

Order Perciformes (L. *perca*, perch + *formis*, having the form of). Fins usually with spines; air bladder without duct; pelvic fins located far forward. Largest order of vertebrates with over 8000 species. Sunfish (*Lepomis*), perch (*Perca*), mackerel (*Scomber*), barracuda (*Sphyraena*).

Class Amphibia (Gk. *amphi*, double + *bios*, life). Skin moist, glandular, and lacking external scales; two pairs of limbs usually present; paired nostrils which open into mouth cavity; skull with two occipital condyles. Three orders and over 2500 species.

Order Gymnophiona (Gk. *gymnos,* naked + *ophioneos,* snake-like). This order is also sometimes called Apoda (Gk. *a,* without + *podos,* feet). Legless, wormlike amphibians; some with dermal scales imbedded in skin. About 50 species. Caecilians (*Icthyopsis*).

Order Urodela (Gk. *ura,* tail + *delos,* visible). This order is also sometimes called Caudata (L. *cauda,* tail). Tail; two pairs of limbs about equal in size—no scales. Over 250 species. Salamander (*Ambystoma*), hellbender (*Cryptobranchus*), mud puppy (*Necturus*), newt (*Diemictylus*).

Order Saliendia (L. *saliens,* leaping). This order is also called Anura (Gk. *a,* without—*ura,* tail). Tail absent, few vertebrae, hind legs adapted for leaping, scales absent. Over 2200 species. Frogs (*Rana*), tree toads (*Hyla*), toads (*Bufa*).

Class Reptilia (L. *repere,* to creep). Skin dry, usually with horny scales; typically with two pairs of limbs, each bearing five toes with claws (limbs absent or reduced in some); skeleton bony with well-developed ribs. Four living orders, over 6000 species.

Order Squamata (L. *squamatus,* scaly). Skin with horny, epidermal scales; quadrate bone movable; male with reversible copulatory organ (hemipenes); anus a transverse slit. Over 3000 species of lizards and 2500 species of snakes. Lizards: Gecko (*Coleonyx*), iguana (*Iguana*), American "chameleon" (*Anolis*), horned toad or lizard (*Phrynosoma*), skink (*Eumeces*), Gila monster (*Heloderma*). Snakes: Anaconda (*Eunectes*), water snake (*Natrix*), black snake (*Coluber*), king snake (*Lampropeltis*), water moccasin (*Agkistrodon*), rattlesnake (*Crotalus*), coral snake (*Micrurus*), cobra (*Naja*).

Order Chelonia (Gk. *chelone,* tortoise). This order is also called Testudinata (L. *testudinatus,* tortoise). Body covered with a bony case made up of a dorsal, rounded carapace and a ventral, flat plastron; jaws lack teeth, covered with horny sheaths; quadrate bone immovable; ribs and some vertebrae fused to carapace; anus a longitudinal slit. Over 300 species. Snapping turtle (*Chelydra*), painted turtle (*Chrysemys*), box turtle (*Terrapene*), land tortoise (*Gopherus*), loggerhead (*Caretta*), soft-shelled turtle (*Trionyx*).

Order Crocodilia (L. *crocodilus,* crocodile). Bony elongate; head large and long with powerful jaws provided with many conical teeth; two pairs of short limbs; digits with claws and webbed; skin very thick and leathery; quadrate bone immovable; anus a longitudinal slit. About 25 species. Crocodile (*Crocodylus*), alligator (*Alligator*).

Order Rhynchocephalia (L. *rhyncho,* snout + Gr. *kephale,* head). Lizardlike animal with a third eye on top of its head; quadrate bone immovable; anus a transverse slit. A single living species, the tuatara of New Zealand (*Sphenodon punctatum*).

Class Aves (L. *avis,* bird). Body covered with feathers; anterior appendages modified as wings; posterior appendages are legs covered with scales; mouth extends as a pointed beak and lacks teeth; body temperature maintained constant—homoiothermal. Twenty-seven orders and over 8000 species.

Order Sphenisciformes (Gk. *spheniscus,* wedge + form). Flightless, forelimbs paddlelike; feet webbed; feathers small and scalelike. Seventeen species. Penguins (*Aptenodytes*).

Order Struthioniformes (Gk. *struthio,* ostrich + form). Flightless; sternum unkeeled; long, strong legs with only two toes; long neck. One species, the ostrich (*Struthio camelus*).

Order Rheiformes (Gk. *struthio,* ostrich + form). Flightless; sternum unkeeled; elongated neck and legs; three toes. Two species. Rhea (*Rhea*).

Order Casuariiformes (Malay, *kasuari* + form). Flightless; sternum unkeeled; three toes on each foot; neck well feathered. Four species. Cassowary (*Casuarius*), emu (*Dromiceius*).

Order Apterygiformes (Gk. *a,* without + *pteryx,* wing + form). Flightless; beak long and slender with nostrils at tip; four toes. Three species. Kiwi (*Apteryx*).

Order Tinamiformes (Carib, *tinamu* + form). Wings short, rounded; tail short. Forty-two species. Tinamou (*Crypturellus*).

Order Gaviiformes (L. *gavia,* sea mew + form). Legs short, located at posterior of body; excellent divers. Four species. Loon (*Gavia*).

Order Podicipediformes (L. *podic-, podex,* rump + *pes,* foot + form). Tail vestigial; legs placed far back on the body; feet lobed; diving birds. Eighteen species. Grebes (*Podiceps*).

Order Procellariiformes (L. *procella,* tempest + form). Tubular nostrils; bill with sheath made of several plates; wings long; feet webbed and lacking hind toe. Seventy-seven species. Albatross (*Diomedea*), petrel (*Procellaria*).

Order Pelicaniformes (L. *pelicamus,* pelican + form). Nostrils reduced or absent; throat pouch present; foot web includes all four toes. Fifty species. Pelican (*Pelecanus*), cormorant (*Phalacrocorax*), booby (*Sula*).

Order Ciconiiformes (L. *ciconia,* stork + form). Long-necked; long-legged wading birds. One hundred and nineteen species. Stork (*Ciconia*), egret (*casmerodius*), heron (*Butorides*).

Order Anseriformes (L. *anser,* goose + form). Aquatic; bill broad with hardened cap at tip; tail short; feet webbed. One hundred and forty-seven species. Mallard duck (*Anas*), goose (*Branta*), swan (*Cygnus*).

Order Falconiformes (L. *falco,* falcon + form). Bill hooked, sharp-edged and stout; feet with sharp, long, curved claws and adapted for seizing prey. Two hundred and seventy-two species. Hawk (*Buteo*), turkey buzzard (*Cathartes*), condor (*Gymnogyps*), osprey (*Pandion*). The bald eagle seen on the United States Seal is in this order—*Haliaeetus leucocephalus.*

Order Galliformes (L. *gallus,* cock + form). Feet adapted for running and scratching; poor fliers; game birds which mostly nest on the ground. Two hundred and forty species. Grouse (*Bonasa*), prairie chicken (*Tympanuchua*), bobwhite quail (*Colinus*), domestic chicken (*Gallus*), turkey (*Meleagris*), pheasant (*Phasianus*).

Order Gruiformes (L. *grus,* crane + form). Long-necked, long-legged marsh birds; some flightless species. One hundred and eighty-six species. Crane (*Grus*), rail (*Rallus*), coot (*Fulica*).

Order Charadriiformes (NL from *Charadrius,* the type genus + form). Shore birds with long legs and webbed feet; dense plumage. Two hundred and ninety-three species. Gulls (*Larus*), terns (*Sterna*), snipe (*Capella*), sandpiper (*Erolia*), plover (*Pluvialis*), puffin (*Fratercula*).

Order Columbiformes (L. *columba*, dove + form). Bill slender and short; large crop which secretes "pigeon milk" to feed young. Three hundred and two species. Pigeon (*Columba*), doze (*Zenaidura*).

Order Psittaeciformes (L. *unreadable* text + form). Beak hooked, stout, narrow and sharp-edged with upper portion movable on bone of skull; two toes in front and two behind with the outer hind toe not reversible; plumage highly colored. Three hundred and sixteen species. Parrot (*Psittacus*), macaw (*Ara*), parakeet (*Myopsitta*).

Order Cuculliformes (L. *cuculus*, cuckoo + form). Two toes in front and two behind with the outer back toe reversible; long tail; old world cuckoo a parasite with female laying her eggs in nests of other birds; American species not parasitic. One hundred and forty-three species. Cuckoo (*Coccyzus*), road-runner (*Geococcyx*).

Order Strigiformes (Gk. *strix*, owl + form). Large head; eyes large and face forward; large ear opening, often with a flaplike covering; beak short, stout and hooked; claws sharp; feet adapted for grasping; largely nocturnal. One hundred and thirty-two species. Barn owl (*Tyto*), horned owl (*Bubo*), snowy owl (*Nyctea*).

Order Caprimulgiformes (L. *caprimulgus*, goatsucker + form). Beak small; mouth wide and often fringed with feathers; legs short; feed on the wing, catching insects at dusk and later. Whippoorwill (*Antrostomus*), nightjar (*Caprimulgus*), night-hawk (*Chordeiles*).

Order Apodiforms (Gk. *apour*, footless + form). Small birds; legs very short and feet very small; hummingbirds with tubular beak and tongue; swifts with small beaks. Three hundred and eighty-seven species. Hummingbird (*Archilochus*), chimney swift (*Chaetura*).

Order Culiiformes (Gk. *kolios*, woodpecker + form). Small, long-tailed birds; all four toes can be directed anteriorly. Six species, confined to Africa. Mouse bird (*Colius*).

Order Trogoniformes (Gk. *trogom,* gnawing + form). Plumage very brilliant; often green; bill stout and short; feet small with first and second toes directed backward. Thirty-five species. Trogon (*Trogon*), quetzal (*Pharomacrus*).

Order Coraciiformes (Gk. *korax,* raven + form). Strong bill, third and fourth toes fused basally; often brilliantly plumed. One hundred and ninety-two species. Kingfisher (*Megaceryle*), hoopoe (*Upupa*).

Order Piciformes (L. *picus,* woodpecker + form). Beak long, stout and pointed; tongue protrusible; tail feathers stiff and pointed. Three hundred and seventy-seven species. Woodpecker (*Dendrocopos*), sapsucker (*Sphyrapicus*), flicker (*Colaptes*), toucan (*Ramphastos*).

Order Passeriformes (L. *passer,* sparrow + form). Three toes in front and one behind, adapted for perching. This order includes more than half of all known species and is made up of four suborders, 69 families, and over 5000 species. Flycatcher (*Tyrannus*), lark (*Otocoris*), swallow (*Hirundo*), crow (*Coreus*), bluejay (*Cyanocitta*), wren (*Troglodytes*), robin (*Turdus*), starling (*Sturnus*), warblers (*Vermivora*), meadowlark (*Sturnella*), cardinal (*Richmondena*), sparrow (*Ammospiza*).

Class Mammalia (L. *mamma,* breast). Body covered with hair, at least in part; skin with several types of glands; toes with claws, nails, or hoofs; females with mammary glands which furnish nourishment to the young. Two subclasses, 18 orders and over 4000 species.

Subclass Prototheria (Gk. *proto,* first + *therion,* beast). No external ear; teeth only in young, adults with horny beak; cloaca present; mammary glands lack nipples; females lack uterus or vagina and lay eggs with soft, pliable shells. One order and five species.

Order Monotremata (Gk. *monos,* single + *tremos,* hole). The single order has the characteristics of the subclass, given above. Duck-billed platypus (*Ornithorhynchus*) and spiny anteater or echidna (*Tachyglossus*).

Subclass Theria (Gk. *therion,* beast). External ears present; teeth in both young and adult; nipples present; uterus and vagina present in female; cloaca usually absent; females viviparous (produce living young). Two infraclasses.

Infraclass Metatheria (Gk. *meta*, after + *therion*, beast). Young born in immature condition and complete development in the marsupium, a ventral pouch. One order and about 250 species, most of which are restricted to Australia.

Order Marsupialia (Gk. *marsypion*, pouch). The single order has the charactistics of the infraclass, given above. Opossum (*Didelphis*), kangaroo (*Macropus*), koala bear (*Phascolarctus*), wombat (*Phascolomis*).

Infraclass Eutheria (Gk. *eu*, true + *therion*, beast). No marsupial pouch; fetus develops entirely within the body of the female and is nourished by way of the placenta. Sixteen orders and almost 4000 species.

Order Insectivora (L. *insectum*, insect + *vorare*, to eat). Small in size; elongated snout; toes with claws and usually five in number; sharp-pointed teeth. About four hundred species. Hedgehog (*Erinaceus*), mole (*Scalopus*), shrew (*Sorex*).

Order Dermoptera (Gk. *derma*, skin + *pteron*, wing). Web of skin between body and the limbs and tail allows the animal to glide through the air. Two species. ''Flying lemur'' (*Galeopithecus*).

Order Chiroptera (Gk. *cheir*, hand + *pteron*, wing). Four digits of forelimb elongated supporting membranous wing which includes the hind limbs and often the tail; capable of true flight; mostly nocturnal. About 900 species. Brown bat (*Myotis*), fruit bat (*Pteropus*), vampire bat (*Desmodus*).

Order Primates (L. *prima*, first). Mostly tree-dwelling forms; head freely movable on the neck; five digits on each limb, usually with nails; thumbs and big toes usually opposable; cranium highly developed in many. About 200 species. Lemur (*Lemur*), galago (*Galago*), tarsier (*Tarsius*), howler monkey (*Alouatta*), marmoset (*Callithrix*), rhesus monkey (*Macaca*), gibbon (*Hylobates*), chimpanzee (*Anthropopithecus*), gorilla (*Gorilla*), and man (*Homo sapiens*).

Order Edentata (L. *edentatus,* toothless). Only molars without enamel, or teeth completely absent; toes clawed. About 30 species. Anteater (*Myrmecophaga*), sloth (*Bradypus*), armadillo (*Dasypus*).

Order Pholidota (Gk. *pholis,* horny scale). Large overlapping horny plates covering the body, with sparse hair between the plates; teeth absent; tongue long and slender. Eight species. Scaly anteater (*Manis*).

Order Lagomorpha (Gk. *lagor,* hare + *morphe,* form). Incisors grow continually, two pairs in upper jaw and one pair in lower; tail short and stubby; no canine teeth; elbow joint unable to rotate. About 60 species. Pika (*Ochotona*), varying hare (*Lepus*), cottontail (*Sylvilagus*).

Order Rodentia (L. *rodare,* to gnaw). Incisors grow continually, one pair per jaw, with enamel on front surfaces only; no canines; elbow rotatable. About 1700 species. Squirrel (*Sciurus*), ground squirrel (*Citellus*), chipmunk (*Tamias*), woodchuck (*Marmota*), pocket gopher (*Geomys*), pocket mouse (*Perognathus*), beaver (*Castor*), deer mouse (*Peromyscus*), house mouse (*Mus*), rat (*Rattus*), porcupine (*Erethizon*), guinea pig (*Cavia*).

Order Cetacea (L. *Cetus,* whale). Medium to very large size; head long and without neck; forelimbs paddlelike, hind limbs absent; nostrils located dorsally; small ear openings; teeth lacking enamel or absent. About 80 species. Sperm whale (*Physeter*), porpoise (*Phocaena*); narwhal (*Monodon*), killer whale (*Orcinus*), right whale (*Balaena*), blue whale (*Balaenoptera*).

Order Carnivora (L. *caro,* flesh + *vorare,* to eat). Clawed toes, usually five and at least four; canines long; teeth pointed. About 280 species. Dog and wolf (*Canis*), fox (*Vulpes*), bear (*Ursus*), raccoon (*Procyon*), weasel (*Mustela*), skunk (*Mephitis*), cat, lion, tiger (*Felis*), sea lion (*Zalophus*), walrus (*Odobenus*).

Order Tubulidentata L. *tubulus,* tube + *dens,* tooth). Long tubular snout; protrusible tongue; teeth lacking enamel; toes with heavy claws. A single species, the aardvark, *Orycteropus afer.*

Order Proboscidea (Gk. *pro,* before + *boscein,* to feed). Huge body; broad ears; pillarlike legs; skin thick; two upper incisors very long (tusks); nose and upper lip extended as long, flexible proboscis. Three species, the African elephant, (*Loxodonta africana*), the Indian elephant, (*Elephas maximus*), and the pigmy elephant of West Africa, *Elephas cyclotis.*

Order Hyracoidea (Gk. *hyrex,* a shrew). Short ears and tail; four toes on forelimb and three on the rear ones. Five species. Coney (*Procavia*).

Order Sirenia (Gk. *seiren,* sea nymph). Hind limbs absent, forelimbs paddlelike; tail with lateral flukes; blunt snout and fleshy lips; no external ears; little hair present. Five species. Manatee (*Trichechus*), dugong (*Halicore*).

Order Perissodactyla (Gk. *perissas,* odd + *dactylos,* toe). Large, long-legged mammals with one or three toes, each provided with a hoof. About 15 species. Horses, asses, and zebras (*Equus*), tapir (*Tapirella*), rhinoceros (*Rhinoceros*).

Order Artiodactyla (Gk. *artios,* even + *dactylos,* toe). Two or four toes, each provided with a hoof; many with antlers or horns. About 170 species. Pig (*Sus*), hippopotamus (*Hippopotamus*), camel (*Camelus*), llama (*Auchema*), deer (*Odocoileus*), wapiti or American elk (*Cercua*), moose (*Alces*), giraffe (*Giraffa*), pronghorn (*Antilocarpa*), sheet (*Ovis*), goat (*Capra*), bison or American buffalo (*Bison*), musk-ox (*Ovibos*), cattle (*Bos*).

Appendix 2

Sample Problems Using
the Proposed Methodology

QUESTION:

The 96 hour LC_{50} for the adult sunfish is 2.5 mg/L in a flow-through system in the lab for CCl_4. Please estimate the MATC for CCl_4 in order to protect the sunfish in Puffer's Pond, Amherst, MA.

ANSWER:

Refer to Barnthouse et al. (1990) (Figure 5.1 and Table 5.1 of the present document, follow pathway for Box 5, Figure 5.1).

1. Problem: Acute → Chronic Extrapolation
   Given: 96 hr. LC_{50} = 2.5 µg/L (Sunfish, CCl_4)
   Want: estimate of MATC

There are four possible acute → chronic logistic extrapolations depending on endpoints (see Table 5.1, extrapolation equations 7–10).

(1) LC_{50} → HATCH EC25
   Y = 1.1(.40) − 1.2
   Y = − .76 [inv log − .76 = .17 µg/L] (best estimate)

Thus, 0.17 µg/L is the best estimate of the concentration of CCl_4 that would result in a HATCH EC25. However, we will now solve

213

for the estimate at the lower 95% prediction interval (CI) using the "prediction interval" term of Barnthouse et al. (1990) (see Table 5.1).

$-.76 - 1.7 = -2.46$ (best estimate—95% PI)
inv log $-2.46 = .0035 \ \mu g/L$ (estimate at lower 95% PI)

(2) $LC_{50} \rightarrow$ MORT1 EC25

$Y = .87(.40) - .87$
$Y = -.522$ [inv log $-.522 = .300 \ \mu g/L$] (best estimate)
$-.522 - 1.5 = -2.022$
inv log $-2.022 = .0095 \ \mu g/L$ (estimate at lower 95% PI)

(3) $LC_{50} \rightarrow$ MORT2 EC25

$Y = 1.0(.40) - .89$
$Y = -.49$ [inv log $-.49 = .32 \ \mu g/L$] (best estimate)
$-.49 - 1.5 = -1.99$
inv log $-1.99 = .010 \ \mu g/L$ (estimate at lower 95% PI). This endpoint is the least sensitive of the four.

(4) $LC_{50} \rightarrow$ EGGS EC25

$Y = 1.1(.40) - 1.89$
$Y = -1.45$ [inv log $- 1.45 = .035 \ \mu g/L$] (best estimate)
$-1.45 - 1.8 = -3.25$
inv log $-3.25 = .00056$ (estimate at lower 95% PI). This endpoint is the most sensitive of the four.

*NOTE: MORT 1 = Mortality of parental fish
MORT 2 = Mortality of larval fish

QUESTION:

An experiment was recently completed assessing the effects of Chemical X on reproductive indices in Salmon; The data on larval survival are given below:

Larval Survival Treatment	Concentration	# of Larvae Survived/ Total Number
#5	1000 μg/L	6/25
#4	500 μg/L	8/25
#3	100 μg/L	18/25
#2	50 μg/L	21/25
#1	10 μg/L	22/25
Control	0.0 μg/L	24/25

Calculate a chronic MATC for the largemouth bass based on these data. Note: The data are chronic, but not on species of choice.

ANSWER:

Step 1—Use logistic regression to plot data and find MORT2 EC25 for Salmon. An estimate of MORT2 EC25 from this data may be 100 μg/L (We will use this estimate to complete the problem)
  Problem: Want MATC of Chemical X for Largemouth Bass
  Assuming the LC_{50} can be estimated from the EC25 (MORT2) we can use 100 μg/L (log of 100 = 2) in Barnthouse et al.'s acute-to-chronic logistic extrapolation for MORT2 EC25-to-LC_{50}.

$$Y = Mx + b$$
$$2 = 1.0 \, (LC_{50}) - 0.89$$
$$2.89 = LC_{50}$$
$$\text{inv log of } 2.89 = 776 \text{ ug/L}$$

Step 2—Since Barnthouse used LC_{50} data to derive his taxonomic extrapolation equations we may use our data-derived LC_{50} value of 2.89 in the Salmoniformes $\rightarrow$ Perciformes extrapolation equation.

$$Y = Mx + b$$
$$2.89 = 0.94 \, (LC_{50}) + 0.33$$
$$\frac{2.89 - 0.33}{0.94} = LC_{50}$$
$$\text{inv log of } 2.72 = 525 \ \mu\text{g/L}$$

QUESTION:

It is necessary to estimate the MATC of chloral hydrate on Perch. No data exist for this compound with Perch. However, this agent was assessed in Salmon and found to be highly toxic with an LC_{50} of 10 $\mu g/L$. These two species have been independently assessed for 62 agents with respect to LC_{50} data in comparable systems (start with Box 6 for Figure 5.1).

Estimate the MATC for Perch for chloral hydrate.

ANSWER:

Two different extrapolations are needed and are linked together. These include:

(a) Acute-to-Chronic Extrapolation
(b) Taxonomic Extrapolation
Given: LC_{50} (Salmon) = 10 $\mu g/L$
Want: Estimate of MATC for Perch

There are three possible approaches to solving this problem:

(1) Perform acute-to-chronic extrapolation and use this estimated value in the taxonomic extrapolation equation.
(2) Perform taxonomic extrapolation and use this estimated value in acute-to-chronic extrapolation.
(3) Derive separate and independent uncertainty factors (UFs) from each equation, multiply them and then divide them into the LC_{50} given.

NOTE: As seen in problem #1 there are four possible acute-to-chronic extrapolation choices (Table 5.1, Equations 7–10). Use LC_{50} → EGGS EC50 in this example because it was the most sensitive endpoint in problem #1.

Approach #1

A. Acute-to-Chronic followed by taxonomic extrapolation. LC_{50}-to-EGGS EC25 (Equation 10 of Table 5.1).

$$Y = 1.1(x) - 1.89 \qquad \text{Prediction interval} = 1.8$$
$$X = \log LC_{50} = \log 10 = 1$$
$$Y = 1.1(1) - 1.89 = 0.79 \text{ (best estimate)}$$
$$Y = -0.79 - 1.8 = -2.59$$

inv log -2.59 = .0026 μg/L (estimate at the lower 95% PI). Use this value (-2.59) in taxonomic extrapolation process.

B. Taxonomic Extrapolation
    Salmoniformes-to-Perciformes (Use Equation 38 from Table 5.1)

$$Y = .94(x) + .33 \qquad \text{Prediction int.} = 1.09$$
$$Y = .94(-2.59) + .33 = -2.10 \text{ (best estimate)}$$
$$-2.10 - 1.09 = -3.19$$

inv log -3.19 = .00064 μg/L (final extrapolated valued (MATC) estimated at the lower 95% PI).

Approach #2

A. Taxonomic extrapolation followed by acute-to-chronic.
    Salmoniformes-to-Perciformes (Equation 38, Table 5.1):

$$Y = .94(x) .33 \qquad \text{Prediction int.} = 1.09$$
$$Y = .94(1) + .33$$
$$Y = 1.27 \text{ (best estimate)}$$
$$1.27 - 1.09 = .18 \text{ (use this value in Part B)}$$

inv log .18 = 1.5 μg/L (estimate at 95% lower PI)

B. LC_{50}-to-EGGS EC25 (Equation 10, Table 5.1)

$$Y = 1.1(x) - 1.89 \qquad \text{Prediction int.} = 1.8$$
$$Y = 1.1(.18) - 1.89$$
$$Y = -.69 - 1.8 = -3.49$$

inv log -3.49 = .00032 μg/L (final extrapolated value (MATC) estimated at the lower 95% PI).

Approach #3

Derive Independent Uncertainty Factors
Acute-to-Chronic UF, LC_{50}-to-EGGS EC25 (UF #1)

$$Y = 1.1(1) - 1.89 \qquad \text{Prediction int. } 1.8$$
$$Y = -.79 - 1.8 = -2.59$$
$$\text{inv log } -2.59 = .0026$$
$$UF = 10/.0026 = 3846.2. \text{ UF is obtained by divid-}$$
ing the LC_{50} by the lower 95% PI

Taxonomic UF Salmoniformes-to-Perciformes (UF #2)

$$Y = .94(x) + .33 \qquad \text{Prediction int.} = 1.09$$
$$Y = .94(1) + .33$$
$$Y = 1.27 - 1.09 = .18$$
$$\text{inv log } .18 = 1.5$$
$$UF = 10/1.5 = 6.67. \text{ This UF is obtained by divid-}$$
ing the LC_{50} by the lower 95% PI

SF #1 and SF #2 are then multipled as follows:
$$(6.67)(3846.2) = 25641.3$$

The product of the multiplication is the total UF and is divided into the LC_{50} value to derive the MATC:

$$10/25641.3 = .00039 \ \mu g/L \text{ (final extrapolated value (MATC)}$$
estimated at the lower 95% PI).

Results of Three Possible Approaches

(1) Acute-to-Chronic Extrapolation Followed by Taxonomic Extrapolation = .00064 $\mu g/L$
(2) Taxonomic Extrapolation Followed by Acute-to-Chronic Extrapolation = .00032 $\mu g/L$
(3) UF Approach = .00039 $\mu g/L$

Which approach is the most appropriate? While there is substantial agreement among the three approaches, there is no absolute assurance which procedure is best at this time. It is possible that each approach is estimating a somewhat different, but substantially overlapping set of phenomena. Further research is needed in this area.

QUESTION:

Life-stage toxicity tests revealed that hatching rate was adversely affected in Rainbow Trout by copper acetate (follow Box 3 of Figure 5.1). The data were as follows:

Hatching Success	# of Eggs Hatched/ Total # of Eggs
8.0 mg/L	20/100
4.0 mg/L	40/100
2.0 mg/L	55/100
1.0 mg/L	70/100
0.0 mg/L	75/100

Estimate the chronic MATC for Rainbow Trout for copper acetate.

ANSWER:

Use appropriate (e.g., logistic) regression to find EC25 HATCH of trout. Here we will estimate $55/100 = 2.0$ $\mu g/L$ as the EC25 HATCH. You may want to stop here if interested in this endpoint or use Barnthouse et al.'s life-stage extrapolations to determine the most sensitive life-stage endpoint.

HATCH EC25-to-EGGS EC25
HATCH EC25-to-MORT1 EC25

HATCH EC25-to-EGGS EC25
$\qquad Y = 6.78(x) + 0.11 \qquad$ Prediction int. $= 1.1$
$\quad \log 2.0 = 0.30$
$\qquad Y = 0.78(.30) + 0.11\ 0.344$ (best estimate)
$\qquad Y = 0.344 - 1.1 = -.756$
inv log $-.756 = .175$ $\mu g/L$ (estimate at the lower 95% PI)

or
HATCH EC25-to-MORT1 EC25
$\qquad Y = 0.80(x) - 0.05 \qquad$ Prediction int. $= 1.1$
$\qquad Y = 0.80(0.30) - 0.05 = 0.19$ (best estimate)

$$Y = 0.19 - 1.1 = -.91$$
$$\text{inv log } -.91 = .123 \ \mu g/L \text{ (estimate at the lower 95\% PI)}$$

Thus, 0.123 $\mu g/L$ is the lowest estimate of MATC at the lower 95% PI.

DEVELOPMENT OF A POPULATION-BASED TRV

Toxicity data was obtained from HLA (1991) for species of concern at the Rocky Mountain Arsenal (RMA) exposed to DDE/DDT and are listed in Table A2.1. This population-based evaluation employs the extrapolation model developed by Van Straalen and Denneman (1989) based on the model of Kooijman (1987). Summary of the symbols used and calculations are presented in Table A2.2.

Table A2.1. Data Available for Calculation of Population-Based TRV

COC	Species of Concern	Study Type	Dose (mg/kg/d)	Test Species
DDE/DDT	Mallard Duck	LOAEL	4	Duck
DDE/DDT	Mouse	LOAEL	35.7	Mouse
DDE/DDT	Prairie Dog	LOAEL	12.1	Rat

Table A2.2. Summary of Symbols and Calculations

Parameter	Symbol	Value
Number of species tested	m	3
Mean of ln(NOAEL)	X_m	0.183
Geometric mean NOAEL	$\overline{\text{NOAEL}}$	1.2
Standard deviation of ln(NOAEL)	S_m	1.095
Fraction of life not protected by HCp[a]	d_1	0.05
Probability of estimating HCp too high	d_2	0.05
Factor[b] dependent on m and d_2	d_m	3.40
Safety factor applicable to mean NOAEL	T	29

[a]HCp = hazardous concentration for p% of the species in a community.
[b]Obtained from Kooijman (1987) Table A2.1.

A safety factor (T) protecting $(100-d_1)\%$ of the species in a community is given by the equation:

$$T = \exp\frac{3S_m d_m}{\pi 2} \ln\frac{1 - d_1}{d_1}$$

The following is a step-by-step procedure for calculation of a safety factor (T) for species of concern described above protecting 95% of the species in the community.

1. Conversion of data from LOAEL to NOAEL involves dividing LOAEL values by a factor of 10 (Table A2.3).
2. The NOAEL values are next converted to natural logarithms (Table A2.3).
3. The mean of the ln(NOAEL)s (X_m), calculated by summing the ln(NOAEL)s and dividing by 3, is determined to be 0.1827. This value is then used to calculate the geometric mean NOAEL (NOAEL) by determining $\exp(X_m)$: $e^{0.1827} = 1.2$.
4. The standard deviation of the ln(NOAEL)s (S_m) is calculated by taking the square root of: $\dfrac{\Sigma(X - X_m)^2}{n - 1}$

   where $X = $ ln(NOAEL)s listed in Table A2.3, $X_m = 0.1827$, and $n = 3$. Therefore, $S_m = 1.09$.

5. The fraction of the species to be protected is arbitrarily chosen to be 0.95. Therefore, the fraction of species *not* protected (d_1) is 0.05.
6. The probability of estimating the hazard concentration for p% of the species too high (d_2) is arbitrarily chosen to be 0.05.
7. The factor (d_m) dependent on m (3) and d_2 (0.05) is obtained from Table A2.1 (Kooijman, 1987) and is 3.40.

The values calculated in steps 1 through 7 are then plugged into the safety factor equation and solved for T:

$$T = \exp\frac{3 \times 1.09 \times 3.40}{(3.14)^2} \ln\frac{1 - 0.05}{0.05}$$

$$T = e^{3.37}$$

$$T = 29$$

Table A2.3.   NOAEL and ln(NOAEL) Values for Test Species Derived from LOAEL Data

Test Species	LOAEL	NOAEL	ln(NOAEL)
Duck	4	0.40	−0.916
Mouse	35.7	3.57	1.273
Rat	12.1	1.21	0.191

This safety factor (T) of 29 is then applied to the geometric mean NOAEL to determine the DDE/DDT exposure that will protect 95% of the species in a community:

$$\frac{\overline{NOAEL}}{T} = \frac{1.2}{29} = 0.04 \text{ mg/kg/d}$$

If 99% of the species in the community were to be protected the safety factor calculation would change as follows:

$$T = \exp \frac{3 \times 1.09 \times 3.40}{(3.14)^2} \ln \frac{1 - 0.01}{0.01}$$

$$T = e^{5.28}$$

$$T = 197$$

When this safety factor (T) of 197 is applied to the geometric mean NOAEL the DDE/DDT exposure that will protect 99% of the species in a community becomes 7.5-fold lower than the 95% exposure limit:

$$\frac{\overline{NOAEL}}{T} = \frac{1.2}{197} = 0.006 \text{ mg/kg/d}$$

If 99.9% of the species in the community were to be protected the safety factor calculation would change as follows:

$$T = \exp \frac{3 \times 1.09 \times 3.40}{(3.14)^2} \ln \frac{1 - 0.001}{0.001}$$

$$T = e^{7.94}$$

$$T = 2815$$

When this safety factor (T) of 2815 is applied to the geometric mean $\overline{NOAEL}$ the DDE/DDT exposure that will protect 99.9% of the species in a community becomes 13.3-fold lower than the 99% exposure limit and 100-fold lower than the 95% exposure limit:

$$\frac{\overline{NOAEL}}{T} = \frac{1.2}{2815} = \text{mg/kg/d}$$

It should be noted that one must be extremely aware of the requirements of the model and its limitations in providing a defensible estimation of a chemical-specific ecosystem TRV.

COMPARISON OF POPULATION-BASED TRV SAFETY FACTORS CALCULATED WITH AND WITHOUT A SENSITIVE SPECIES

Arbitrary toxicity data (Tables A2.4 and A2.5) were created for 7 and 5 species models for comparison of population-based TRV safety factors calculated with and without inclusion of a sensitive species data (Table A2.8). The calculations employ the extrapolation model developed by van Straalen and Denneman (1989) based on the model of Kooijman. Summary of the symbols used and calculations are presented in Tables A2.6 and A2.7.

Table A2.4. Arbitrary Data on 7 Species for Calculation of Population-Based TRV Safety Factors

Species	NOAEL	ln(NOAEL)
1	20	2.99
2	35	3.55
3	15	2.71
4	22	3.09
5	16	2.77
6	23	3.14
7	0.001	−6.91

Table A2.5. **Arbitrary Data on 5 Species for Calculation of Population-Based TRV Safety Factors**

Species	NOAEL	In(NOAEL)
1	20	2.99
2	35	3.55
3	15	2.71
4	22	3.09
5	0.001	−6.91

Table A2.6. **Summary of Symbols and Calculations for 7 Species Model ± Sensitive Species**

Parameter	Symbol	95% Protection		99% Protection	
		+ Sensitive Species	− Sensitive Species	+ Sensitive Species	− Sensitive Species
Number of species tested	m	7	6	7	6
Mean of In(NOAEL)	X_m	1.62	3.04	1.62	3.04
Standard deviation of In(NOAEL)	S_m	2.82	0.30	2.82	0.30
Fraction of life not protected by HCp[a]	d_1	0.05	0.05	0.01	0.01
Probability of estimating HCp too high	d_2	0.05	0.05	0.01	0.01
Factor[b] dependent on m and d_2	d_m	2.82	2.93	3.52	3.74

[a]HCp = hazardous concentration for p% of the species in a community.
[b]Obtained from Kooijman (1987) Table 1.

The data indicate that the presence of the highly sensitive (or perhaps endangered species) has a profound impact on the derivation of the population-based TRV safety factor (T). Likewise, the presence of the sensitive species on the estimation of the $T_{95\%}$ and $T_{99\%}$ was striking in the present examples, while only a modest impact was observed when the sensitive species was omitted.

Table A2.7.   Summary of Symbols and Calculations for 5 Species Model $\pm$ Sensitive Species

Parameter	Symbol	95% Protection		99% Protection	
		+ Sensitive Species	– Sensitive Species	+ Sensitive Species	– Sensitive Species
Number of species tested	m	5	4	5	4
Mean of ln(NOAEL)	X_m	1.09	3.09	1.09	3.09
Standard deviation of ln(NOAEL)	S_m	4.48	0.35	4.48	0.35
Fraction of life not protected by HCp[a]	d_1	0.05	0.05	0.01	0.01
Probability of estimating HCp too high	d_2	0.05	0.05	0.01	0.01
Factor[b] dependent on m and d_2	d_m	3.06	3.22	3.99	4.25

[a]HCp = hazardous concentration for p% of the species in a community.
[b]Obtained from Kooijman (1987) Table 1.

Table A2.8.   Comparison of Population-Based TRV Safety Factors (T) Calculated for 7 and 5 Species Models With and Without Sensitive Species

# Species	$T_{95\%}$		$T_{99\%}$	
	+ Sensitive Species	– Sensitive Species	+ Sensitive Species	– Sensitive Species
7	13,360	2.2	1.2×10^8	4.8
5	211,039	2.7	7.0×10^{10}	7.9

REFERENCES

Kooijman, S. A. L. M. "A Safety Factor for LC_{50} Values Allowing for Differences in Sensitivity Among Species," *Wat. Res.* 21:269–276 (1987).

Van Straalen, N. M., and C. A. J. Denneman. "Ecotoxicological Evaluation of Soil Quality Criteria," *Ecotox. Env. Safety* 18:241–251 (1989).

Appendix 3

Aquatic Toxicology Mixtures

A. INTRODUCTION

Exposure to multiple chemical contaminants in the aquatic environment has been a concern for over 50 years (Southgate, 1932; Bergstrom and Vallin, 1937; Cherkinsky, 1957; Friedland and Rubleva, 1958; Bucksteeg, 1961; Calamari and Alabaster, 1980; FAO, 1980; Marking, 1985; Lloyd, 1987). However, the first major synthesis of the mixtures problem for the aquatic environment was published by Sprague (1970).

In this widely cited paper, Sprague (1970) functionally linked research efforts in the field of aquatic toxicology with previous efforts and advances in the area of drug and insecticide toxicology, where considerable theoretical and applied research in mixtures had been conducted. Sprague (1970) recommended the adoption of the chemical interaction scheme of Gaddum (1948) for its applications in predicting the responses of fish exposed to multiple toxicants. The approach of Gaddum (1948) was a concentration addition methodology that was visually represented by the isobologram and that had been widely adopted and applied by pharmacologists and toxicologists, especially those involved with assessing toxicity from pesticides.

B. CONCENTRATION ADDITION AND THE TOXIC UNIT CONCEPT

The application of the concentration addition methodology to the aquatic environment for predicting joint toxicity, summarized by Brown (1968), has historically been based on incipient LC_{50} values of contaminants or approximations of 48- or 96-hr LC_{50}s. In operational terms, the strength of a given toxicant was expressed as a fraction or proportion of its lethal threshold concentration. If this ratio exceeds a value of 1.0, more than half of a group of fish would be predicted to be killed by this toxicant. If the ratio were less than 1.0, then less than half the fish would be killed. Sprague and Ramsay (1965) referred to the numbers of this scale as toxic units (TU), based on the earlier suggestion of Bergstrom and Vallin (1937), with 1.0 TU equal to the incipient LC_{50}. The quantification of toxic units for any toxicant is determined as follows:

$$\text{toxic units} = \frac{\text{actual concentration in solution}}{\text{lethal threshold concentration}}$$

Sprague (1970) indicated that the TU for each component of the mixture of concern should be calculated, and these values may be summed since they are all in the same units. Solutions with toxic unit values of ≥ 1.0 are predicted to be lethal, while solutions with toxic unit values of 5–10 are expected to kill the fish in several hours. It should be emphasized that toxic units are species- and environmental condition-specific values.

This toxic unit approach of Brown and numerous others is similar to, and essentially based on, the isobole theory of joint interactions. In 1975, however, Marking and Dawson extended the methods, concepts, and terminology related to additive toxicity to derive a "quantitative index for the toxicity of mixtures of chemicals in water." As with previous attempts, the approach of Marking and Dawson (1975) was also founded on the isobole theory, and it involved the use of the toxic unit concept to essentially sum the action of various components of a mixture. The method used an additivity formula:

$$\frac{Am}{Ai} + \frac{Bm}{Bi} = S$$

where A and B are chemicals, i and m are the toxicities (LC_{50}s) of A and B individually (i) and in a mixture (m), and S is the sum of the responses.

Marking and Dawson (1975) extended this approach to develop a system whereby additive, greater-than-additive, and less-than-additive effects are represented by zero, positive, and negative values, respectively. In the previous formula, greater than additivity would yield a negative number, and less than additivity would yield a positive number. This modification was accomplished by assigning zero as the reference point for simple additive toxicity and establishing linearity for greater-than-additive and less-than-additive values. The zero reference point was obtained by subtracting 1.0 from S, while linearity was achieved by employing the reciprocal of values of S that were less than 1.0—that is, $1/S - 1$. Index values for less-than-additive toxicity are obtained by multiplying values of S that were greater than 1.0 by -1 to make them negative, and a zero reference point was generated by adding 1.0 to this negative value $[S(-1) + 1]$. A sum (S) of 1.0 leads to an index value of zero regardless of which procedure is used and therefore indicates a simple additive response.

Marking (1985) provided an example of the additive toxicity index by citing data of Douderoff (1952), who assessed the toxic activity of mixtures of zinc and copper in fathead minnows using 8-hr EC_{50} values. The data were

$$Zn\ m = 1\ mg/L$$

$$Zn\ i = 8\ mg/L$$

$$Cu\ m = 0.025\ mg/L$$

$$Cu\ i = 0.2\ mg/L$$

$$\frac{Zn\ m}{Zn\ i} + \frac{Cu\ m}{Cu\ i} = S$$

$$\frac{1.0}{8.0} + \frac{0.025}{0.20} = 0.250$$

Additivity Index $= 1/S - 1.0 = 1/0.250 - 1 = 3$

Other examples are provided in Table A3.1. The significance of index values rests with their interpretation. A number of investigators have arbitrarily selected ranges of S values to identify toxicity that was greater than or less than additive. Using a nonlinear scale where a value of 1.0 equals additive effects, Kobayashi (1978) recommended values of S between 0.5 and 2.0 to represent simple

Table A3.1.   Toxicity or Efficacy of Chemicals Applied Individually and in Combination Against Fishes and the Calculated Additive Index

Chemical Mixtures and Toxic Units	96-hr LC_{50} or EC_{50} of Chemicals		Additive Index
	Individually	In Combination	
Zinc[a] (mg/L)	8.0	1.0	
and			3.00
Copper [a] (mg/L)	0.2	0.025	
Zinc (mg/L)	4.2	3.90	
and			−1.37
Cyanide (mg/L)	0.18	0.26	
Antimycin (mg/L)	0.032	0.027	
and			−0.39
Rotenone (mg/L)	57.0	31.0	
MS-222 (mg/L)	80	30	
and			0.29
$CdSO_4$ (mg/L)	25	10	
Malachite green (mg/L)	0.2	0.05	
and			0.83
Formalin[b] (mg/L)	50	15	

Source: Marking and Dawson (1975).
[a]An 8-hr time response, based on survival rather than LC_{50}.
[b]Concentrations effective against parasites.

additive effects. This is comparable to the so-called envelope of additivity. Marking and Dawson (1975) offered a more objective procedure for judging whether the additive index values were different from zero (i.e., additive toxicity) by substituting the 95% confidence intervals for the LC_{50}s into the additive index formula. Mixtures that overlapped zero were deemed to be only additive in toxicity, while those that did not overlap zero were either greater- or less-than-additive toxicity.

Marking (1977) attempted to enhance the meaning of index values by creating a new term, the magnification factor, that describes the magnitude of additive toxicity. More specifically, if the toxicity of a mixture were enhanced twofold (i.e., magnification factor = 2) over predicted simple additive toxicity, then the additive index would be equal to 1.0. Thus, the magnification factor is derived simply by adding 1.0 to the numerical index value. For less-than-additive values, the factor is reciprocated; for example, when an index value is 9, the magnification factor is 10, and when the index value is −9, the magnification factor is 1/10.

The TU method has been widely employed in the aquatic toxicology literature (Alabaster and Lloyd, 1980, 1982; Marking, 1985). In 1980 the European Inland Fisheries Advisory Commission (EIFAC) (FAO, 1980) published an analysis of 76 experiments on the toxicity to fish of mixtures of common pollutants most often in pairs at equitoxic concentrations. The contaminants studied included ammonia, phenol, copper, zinc, cadmium, nickel, chromium, and mercury, which were selected to represent major toxic constituents of sewage effluents and industrial wastes. The findings indicated that the toxicity of the mixture ranged from 0.4 to 26 times the value predicted using the TU method. The vast majority (87%) of the tested joint toxicity values were within 0.5 and 1.5 times the predicted toxicity value, with the median value 0.95 times the predicted toxicity (Table A3.2). Based on these findings, Lloyd (1982) concluded that joint action of actual toxic mixtures of common contaminants in the aquatic environment are likely to be additive. This conclusion was consistent with the results of a comparable assessment of experimental data on mixtures of pesticides and other agents (e.g., surfactants, synergist antibiotics). The actual responses were somewhat greater than predicted, with the median value being 1.3 times greater than the predicted value.

The most controversial observation that emerged from the synthesis of the large data set was that a case could be made to support the premise that where the concentration of a pollutant was less than 0.2 TU, it did not contribute to the toxicity of a mixture. However, Lloyd (1982) has argued that the individual experiments themselves were generally not designed to provide the precision needed to properly assess this hypothesis. However, in the case of mixture studies with phenol, a strong database indicates that concentrations greater than 0.3 TU were additive, while concentrations between 0.1 and 0.3 TU were less than additive. Of considerable interest was the observation that concentrations lower than 0.1 TU were antagonistic; that is, they diminished the toxicity of the other contaminants in the mixture (Figure A3.1). Lloyd (1982) argued that this may represent an example of hormesis, which is the stimulation of biological activity by low concentrations of toxic substances (Stebbing, 1982). Support for the hypothesis that low levels of toxic substances are not likely to achieve predicted

Table A3.2. Laboratory Data on the Joint Action of Mixtures of Toxicants on Fish

Toxicants	Species	Exposure Period and Response	Ratio of Toxicant EC_{50}	Joint Action	Multiple of Additive Joint Action
Ammonia + phenol	Rainbow trout	Threshold LC_{50}	1:1 to 1:2	Additive	0.7 to 1.0
	Fathead minnow	24-hr LC_{50}	1.0:0.1 & 0.3	Less than additive	0.7
	Fathead minnow	24-hr LC_{50}	1.0:0.3 & 0.7	Additive	1.0 to 1.2
Ammonia + cyanide	Rainbow trout	96-hr LC_{50}	1:1	Additive	1.2
		30-day EC_{50} (growth)	1.0:0.7	Less than additive	0.6
Ammonia + copper	Rainbow trout	48-hr LC_{50}	1:1	Additive	1.0
		LC_{25}	1:1	More than additive	1.2
		LC_{10}	1:1	More than additive	1.4
Ammonia + zinc	Rainbow trout	Threshold LC_{50}	1.0:0.5	Additive	1.0
		Hard water	1:2	Additive	1.0
		Soft water	1:1	Less than additive	0.8
Ammonia + phenol + zinc	Rainbow trout	48-hr LC_{50}	1.0:1.0:0.5	Additive	1.0
			1:7:1	More than additive	1.2
			1:1:6	Less than additive	0.7
			1.0:0.2:0.1	Additive	0.9
Ammonia + phenol + sulphide	Fathead minnow	24-hr LC_{50}	1.0:0.1:1.1	Less than additive	0.5
			1.0:0.02:0.1	Less than additive	0.6
			1.0:0.7:2.0	Less than additive	0.6
			1.0:0.3:0.3	Less than additive	0.8

continued

Table A3.2. *Continued*

Toxicants	Species	Exposure Period and Response	Ratio of Toxicant EC_{50}	Joint Action	Multiple of Additive Joint Action
Ammonia + sulphide	Fathead minnow	24-hr LC_{50}	1.0:2.2	Less than additive	0.6
			1.0:0.3	Less than additive	0.6
			1:11	Less than additive	0.8
			1.0:1.4	Less than additive	0.8
Ammonia + nitrate	Guppy	72-hr LC_{50}	1.0:>0.55	Additive	<1.1
			1.0:<0.35	Less than additive	>0.7
Phenol + copper	Rainbow trout	48-hr LC_{50}	1:1	Less than additive	0.85
Phenol + copper + zinc	Rainbow trout	48-hr LC_{50}	1:1:1	Less than additive	0.9
Phenol + sulphide	Fathead minnow	24-hr LC_{50}	1:3.5	Less than additive	0.8
			1:11	Less than additive	0.8
			1:3	Less than additive	0.7
			1:1	Additive	0.9
Cyanide + zinc	Bluegill	96-hr LC_{50}	1:1	Less than additive	0.4
	Fathead minnow	96-hr LC_{50}		More than additive	1.4
		30-day EC_{50} (growth)	1.0:0.6	Less than additive	0.4

continued

Table A3.2. *Continued*

Toxicants	Species	Exposure Period and Response	Ratio of Toxicant EC_{50}	Joint Action	Multiple of Additive Joint Action
Cyanide + chromium	Fathead minnow	96-hr LC_{50}	1:1	Less than additive	0.8
		30-day EC_{50} (growth)	1.0:0.8	Less than additive	0.6 to 0.8
			1.0:0.1	Less than additive	
Copper + zinc	Rainbow trout	3-day LC_{50} (hard water)	1:1	Additive	1.0
		7-day LC_{50} (soft water)	1:1	Additive	1.0
	Atlantic salmon	7-day LC_{50} (soft water)	1:1	Additive	1.0
	Longfin dace	96-hr LC_{50}	1:0.75	More than additive	1.2
Copper + zinc + nickel	Rainbow trout	48-hr LC_{50}	1:1:1	Additive	0.7
	Rainbow trout	Non-specified	1:1:1	Additive	(0.4 to 1.2)
Copper + nickel	Guppy	7-day EC_{50} (growth, restricted ration)	?	Slightly more than additive	?
		7-day EC_{50} (growth unrestricted ration)		Additive	

continued

Table A3.2. *Continued*

Toxicants	Species	Exposure Period and Response	Ratio of Toxicant EC_{50}	Joint Action	Multiple of Additive Joint Action
Copper + cadmium	Mummichog (salinity 20%)	Effect on lateral line	?	Less than additive	?
	(salinity 20%)	96-hr LC_{50}	Various	More than additive	?
	Zebrafish	96-hr LC_{50}	?	More than additive	2.0
Copper + zinc + cadmium	Mummichog (salinity 20%)	96-hr LC_{50}	Various	More than additive	?
	Fathead minnow	96-hr LC_{50} (Cu) Threshold LC_{50} (Cd and Zn) 12½ months	1.0:0.1:1.2	More than additive	1.3
		EC_{50}	1.0:0.1:1.2	Varied, depending on response measured	
		90% red, in no eggs	1.0:0.1:1.2	More than additive	>1.6
		50% red, in no eggs	1.0:0.2:1.2	Less than additive	<0.9
Copper + mercury	Rainbow trout	96-hr LC_{50}	?	Additive	?

continued

Table A3.2. *Continued*

Toxicants	Species	Exposure Period and Response	Ratio of Toxicant EC_{50}	Joint Action	Multiple of Additive Joint Action
Copper + mercury (methyl)	Blue gourami	96-hr LC_{50}	1.0:0.25 to 1.0:1.15	Less than additive	?
Copper + manganese	Longfin dace	96-hr LC_{50}		Less than additive	0.67
Copper + surfactant ABS	Goldfish	96-hr LC_{50}	1.0:0.1 1.0:0.6	More than additive More than additive	1.3 2.1
LAS	Rainbow trout	96-hr LC_{50}	1:1	More than additive	1.3
+ nonyl phenyl ethoxylate	Rainbow trout	96-hr LC_{50}	1:1	Less than additive	0.8
Copper + paraquat	*Poecilia mexicana*	24-hr LC_{50}	1:0.25	More than additive	1.3
Zinc + cadmium				Various	
Zinc + detergent (55% ABS)	Rainbow trout	72-hr LC_{50}	1:3.8	Additive	0.9

continued

Table A3.2. *Continued*

Toxicants	Species	Exposure Period and Response	Ratio of Toxicant EC_{50}	Joint Action	Multiple of Additive Joint Action
Cadmium + chromium + nickel	Rainbow trout	3-month physiological	?	Less than additive	?
Cadmium + mercury	?	96-hr LC_{50} 10-day LC_{50}	? ?	Additive More than additive	1.0 2.0
Mercury + surfactant (LAS)	Rainbow trout	96-hr LC_{50}	Various	Additive	1.1
Chromium + nickel	Rainbow trout Rainbow trout Rainbow trout	96-hr LC_{50} 10-week LC_{50} 10-week LC_{50}	1:0.3 1:0.3 1:1	Additive More than additive More than additive	0.9 13 21
Chromium + surfactant	Alborella	96-hr LC_{50}	1:17 to 1.0:0.8	Less than additive	0.6 to 0.7

Source: Food and Agriculture Organization of the UN (1980).

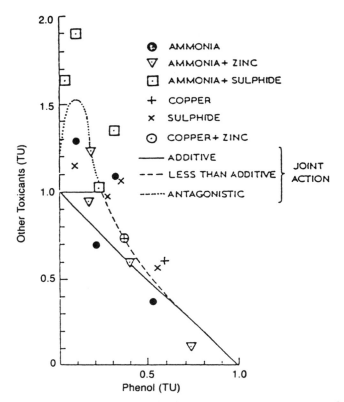

Figure A3.1. Toxic units of phenol and other toxicants in mixtures producing a given lethal response in fish. From Food and Agriculture Organization of the U.N. (1980).

toxicity via the additivity model is seen in studies conducted on the effects of mixtures on sublethal responses of fish (e.g., growth and reproduction). According to Lloyd (1982), such findings suggest that the contaminants were considerably less than additive in their combined responses, even though additive when mortality was the endpoint. Lloyd argued that these findings supported the notion that pollutant concentrations less than 0.2 TU that approach the no-observed-effect concentration (NOEC) may not contribute to the toxicity of a mixture.

The premise that exposure to mixtures of aquatic contaminants at ≤0.1 TU is not significant was challenged by a series of experiments of Konemann (1979, 1980), who assessed the toxicity of guppies to a range of 61 organic chemicals including aromatics, chlorinated aromatics, anilines, and glycerols. A mixture of three chlorobenzenes and another of ten chlorobenzenes were slightly less than additive at equitoxic concentrations. In the latter experiment the investigators employed individual concentrations of 0.1 TU, thereby indicating an additive response at a low TU value. Even more impressive was an experiment involving 50 chloroaromatics, chloroalkanes, glycol derivatives, and related compounds, all with the same quantitative structure-activity relationships (QSAR); at equitoxic concentrations (1/50 TU) of the 50 agents, a toxic unit summation of 1.0 was observed, thereby supporting an additivity relationship. Comparable findings were also seen with a mixture of phenol and ten chlorophenols.

Studies by Hermens and Leeuwaugh (1982) assessing the effects of mixtures of equitoxic concentrations of organic compounds, including pesticides, based on 14-day LC_{50} values did not offer validation of either perspective since additivity and less than additivity were observed with various mixture studies involving large numbers of equitoxic agents.

The TU methodology has been applied to complex mixtures in laboratory settings as well as to field studies on polluted rivers. The goal of such research was to assist in the development of water quality standards and assess whether concern over joint effects needed to be taken into consideration. Unfortunately, as Lloyd (1982) pointed out, much of the field data are equivocal as a result of a number of uncontrolled variables. For example, the toxicity of multiple agents may be affected by the concentrations of dissolved oxygen and suspended solids and these are often site-specific. Also, the dynamic nature of flowing streams, with episodic pollutant discharges of variable durations, may also confound predictions that assume a continuous low level of contaminant concentration. Despite the equivocal nature of the field studies, the accumulated findings from the laboratory and field settings led Lloyd (1982) to conclude that "markedly more than additive effects do not occur even when lethal responses of fish are measured. Therefore, when considering water quality standards for mixtures of the common pollutants, allowance need not be made for synergistic effects." Consistent with this viewpoint is the approach of the U.S. EPA (NAS, 1972), which suggests that concentrations lower than 0.2 of the No Observed Effect Concentration

(NOEC) do not contribute to the toxicity of a mixture. This decision was based on professional judgment rather than from experimental findings.

The approach of summing TUs of respective components of a complex mixture represents a crude approach in the estimation of binary or complex mixture exposures. In contrast to mammalian systems, where an effort is made to identify similar mechanisms of toxicity in order to differentiate whether one uses a concentration-vs response-additive approach, the TU aquatic approach has inherently assumed concentration additivity without satisfying its basic assumptions. Consequently it is not unexpected that inconsistencies in the literature exist concerning the validity of summing components with small fractional TU values. Judgments, therefore, that TUs lower than 0.2 of the NOEC should not contribute to the toxicity of a mixture need to be based on both pharmacokinetic and toxicodynamic understandings of the agents in question. At the present time this judgment represents one where the complex mixtures studied have often involved a range of agents that act via unspecified mechanisms on a range of different organ systems. Thus, for this judgment to be valid, the mixture of agents studied should not be acting through similar mechanisms (concentration addition) nor affecting similar organs (or endpoints) via different mechanisms (response addition).

C. DISCUSSION

The attempts to address the challenge of mixtures to aquatic organisms, especially fish, have been reviewed. It has been seen that the approaches employed over several decades by fish toxicologists have been those of simple additivity, following the initial lead of such pioneers as Loewe (1928), Gaddum (1948), and Finney (1942). The predominant endpoint that has been studied has been the 96-hr LC_{50}. Consequently the major emphasis has been in the area of acutely toxic responses. This approach, which has been primarily descriptive in nature ("kill them and count them"), has markedly limited the predictive understandings of how mixtures act within fish models. This is principally the result of the role that the scientific community had allocated to fish in the general environmental arena. Regulatory agencies have not been sufficiently interested in the occurrence and causes of chronic diseases, such as chemically induced cancers, in fish. They have been principally interested in whether there had been fish kills. Only within the last decade has there been a gradual recognition that fish develop a wide variety of environmentally induced chronic diseases and that these models could serve not only as sentinels but also as predictive models used in the process of cross-species extrapolation. Consequently there has been very limited mechanistic insight offered at present by fish toxicologists to the emerging field of the toxicology of mixtures, even though the initial studies on mixture toxicology in fish go back over 50 years. However, this situation is likely to rapidly change since aggressive new research programs, principally by various U.S. agencies (e.g., NCI, EPA, DOD), have begun to consider the predictive utility of fish species in terms of both qualitative and quantitative human responses to environmental agents.

Despite the above-mentioned limitations there have been a number of attempts by fish toxicologists to influence the development of terminology and to extend the manner in which traditional descriptive procedures have been used. These are worthy of detailed consideration and analysis.

As noted earlier, a term used by fish toxicologists in the study of mixtures, but not by any other group, is the *toxic unit*. This term is simply a descriptive expression of a specific type of toxic or exposure ratio. It has been used by fish toxicologists to describe the ratio of a concentration in water to that of the 96-hr LC_{50} value. Regulatory toxicologists have used ratios such as the concentration in drinking water over an acceptable exposure standard. While the concept is the same, it was the first time that a group has given it a formal name.

Several other new terms proposed by Marking over the past decade, including the additivity index and the magnification factor, deserve further discussion. The additivity index is again another summation method used to determine the extent of interaction. The only initial difference is that the numerator and denominator are such that greater-than-additive responses received a fractional (that is, less than 1) value, while less-than-additive responses have values greater than 1. While this is simple and straightforward, the final values are counterintuitive since one has a natural inclination to associate a greater-than-additive response with a larger number. Consequently, Marking developed a mathematical manipulation to convert what he called nonlinear toxic contributions to linearity. In practical terms what this does is to convert a fractional value into a number greater than one, and an initially >1 value (for less-than-additive responses) into a negative number. This now appears to make intuitive sense. However, this mathematical manipulation offers neither mechanistic insight nor enhanced quantitative insight of the degree of interaction. A similar judgment can be made about the magnification factor. It represents another term designed to help toxicologists in dealing with the challenge of mixtures, but it has very limited significance. It simply tries to describe the magnitude of additive toxicity and should not be interpreted beyond that point. Once again, the approach to convert to linearity and to use the magnification factor are appropriate mathematical techniques but offer limited developmental progress.

Despite these criticisms it is important to note that researchers such as Marking and colleagues have attempted to relate their approaches and findings to those employed in the field of mammalian toxicology (Smyth et al., 1969; Keplinger and Deichmann, 1967; Carpenter et al., 1961). These approaches by mammalian toxicologists were essentially limited modifications of the original isobolic methods or extensions of the statistical approaches of Finney. For example, Smyth et al. (1969) normalized the values derived from Finney's (1942) equation with a frequency distribution curve and adjusted the values to indicate additive toxicity with a zero. It is likely, therefore, that over the next decade there will be a markedly greater integration of aquatic and mammalian toxicological research. The field of multiple chemical interactions, while being preadapted to take advantage of such likely progress, will also be a significant contributor to it as well.

REFERENCES

Alabaster, J. S., and R. Lloyd, Eds. (1980). *Water Quality Criteria for Freshwater Fish*, 1st ed. Butterworths, London.

Alabaster, J. S., and R. Lloyd, Eds. (1982). *Water Quality Criteria for Freshwater Fish*, 2nd ed. Butterworths, London.

Bergstrom, H., and S. Vallin. (1937). Vattenfororening genom aulopps-vattnet fran sulfatcellulosafabriker (Water pollution due to waste from sulfate pulp factories). Sweden, Kungl. Lantbruksstyrelsen, Meddelanden fran statens undersoknings-och forsoksanstalt for sotvattensfisket, no. 13. Fish. Res. Bd. Can., translation series no. 582.

Brown, V. M. (1968). The calculation of the acute toxicity of mixtures of poisons to rainbow trout. *Water Research* 2:723–733.

Bucksteeg, W. (1961). Teste zur Beurteilung von Abwassern. *Stadtehygiene* 12:180–184.

Calamari, D., and J. S. Alabaster. (1980). An approach to theoretical models in evaluating the effects of mixtures of toxicants in the aquatic environment. *Chemosphere* 9:533–538.

Carpenter, C. P., C. S. Weil, P. E. Palm, M. W. Woodside, and J. H. Nair, Jr. (1961). Mammalian toxicity of 1-naphthyl-N-methyl-carbamate (Sevin insecticide). *J. Agric. Food Chem.* 9:30–39.

Cherkinsky, S. N. (1957). The theoretical basis of hygienic standardization of simultaneous pollution of water courses with several harmful substances (English summary). *Gig. Sanit.* 22(8):3–9.

Douderoff, P. (1952). Some recent developments in study of toxic industrial wastes. *Proc. Ann. Pac. Northwest Ind. Waste Conf.* 4:1–21.

Finney, D. J. (1942). The analysis of toxicity tests on mixtures of poisons. *Ann Appl. Biol.* 29:82–94.

Food and Agriculture Organization (FAO) of the U.N. (1980). Water quality criteria for European freshwater fish. Report on combined effects on freshwater fish and other aquatic life of mixtures of toxicants in water. European Inland Fisheries Advisory Commission (EIFAC) technical paper no. 37.

Friedland, S. A., and Rubleva, M. N. (1958). The problem of hygienic standards for water simultaneously polluted with several harmful substances (English summary). *Gig. Sanit.* 23:12–16.

Gaddum, J. H. (1948). *Pharmacology*, 3rd ed. O.U.P., London.

Herbert, D. W. M., and D. S. Shurben. (1964). The toxicity to fish of mixtures of poisons. I. Salts of ammonia and zinc. *Ann. Appl. Biol.* 53:33–41.

Hermens, J., and P. Leeuwaugh. (1982). Joint toxicity of mixtures of 8 and 24 chemicals to the guppy (*Poecilia reticulata*). *Ecotoxicol. Environ. Saf.* 6:302–310.

Keplinger, M. L., and W. B. Deichmann. (1967). Acute toxicity of combinations of pesticides. *Toxicol. Appl. Pharmacol.* 10:586–595.

Kobayashi, S. (1978). Synergism in pesticide toxicity. *J. Med. Soc. Toho Univ.* 25:616–634.

Konemann, W. H. (1979). Quantitative structure-activity relationships for kinetics and toxicity of aquatic pollutants and their mixtures in fish. PhD thesis. University of Utrecht, Utrecht, Netherlands.

Konemann, W. H. (1980). Structure-activity relationships and additivity in fish toxicities of environmental pollutants. *Ecotoxicol. Environ. Saf.* 4:415–421.

Lloyd, R., (1982). The toxicity of mixtures of chemicals to fish: an overview of European laboratory and field experience. *Environmental Hazard Assessment of Effluents*, H. L. Bergman, R. A. Kimerle, and A. W. Maki, Eds. John Wiley and Sons, New York.

Lloyd, R. (1987). Special tests in aquatic toxicity for chemical mixtures: interactions and modification of response by variation of physicochemical conditions. In *Methods for Assessing the Effects of Mixtures of Chemicals,* V. B. Vouk, G. C. Butler, A. C. Upton, D. V. Parke, and S. C. Asher, Eds. John Wiley and Sons, New York, pp. 491-507.

Loewe, S. (1982). Die quantitatiren probleme der pharmakologie. *Ergebnisse der Physiologie.* 27:47.

Marking, L. L. (1977). Method for assessing additive toxicity of chemical mixtures. In *Aquatic Toxicology and Hazard Evaluation,* F. L. Mayer, and J. L. Hamelink, Eds. American Society for Testing and Materials, Philadelphia, pp. 99-108.

Marking, L. L. (1985). Toxicity of chemical mixtures. In *Fundamentals of Aquatic Toxicology,* G. M. Rand and S. R. Petrocelli, Eds. Hemisphere Publishing, Washington, DC, pp. 164-176.

Marking, L. L., and V. K. Dawson. (1975). Method for assessment of toxicity or efficacy of mixtures of chemicals. *U.S. Fish Wildl. Serv. Invest. Fish Control* 67:1-8.

National Academy of Sciences and National Academy of Engineering. (1972). Water Quality Criteria 1972. EPA-R3-73-033. U.S. Environmental Protection Agency, Washington, DC.

Smyth, H. F., C. S. Weil, J. S. West, and C. P. Carpenter. (1969). An exploration of joint toxic action: Twenty-seven industrial chemicals intubated in rats in all possible pairs. *Toxicol. Appl. Pharmacol.* 14:340-347.

Smyth, H. F., C. S. Weil, J. S. West, and C. P. Carpenter. (1969). An exploration of joint toxic action. II. Equitoxic versus equivolume mixtures. *Toxicol. Appl. Pharmacol.* 17:498-503.

Southgate, B. A. (1932). The toxicity of mixtures of poisons. *J. Pharmacol.* 5:639-648.

Sprague, J. B. (1970). Measurement of pollutant toxicity to fish. II. Utilizing and applying bioassay results. *Water Res.* 4:3-32.

Sprague, J. B., and B. A. Ramsay. (1965). Lethal levels of mixed copper-zinc solutions for juvenile salmon. *J. Fish. Res. Bd. Can.* 22:425-432.

Stebbing, A. R. J. (1982). Hormesis: the stimulation of growth by low levels of inhibitors. *Sci. Total Environ.* 22:213-234.

List of Acronyms

ACR	acute to chronic ratios
AET	apparent effects threshold
AF	application factor
BCF	bioconcentration factor
BMF	biomagnification factor
CNS	central nervous system
COC	contaminants/chemicals of concern
DOD	Department of Defense
DDT	dichlorodiphenyltrichloroethane
EA	endangerment assessment
EP	equilibrium partitioning
EPA	Environmental Protection Agency
EXAMS	exposure analysis modeling system
FCV	final chronic value
FEL	frank effect level
FM	fathead minnow
FOAM	Fate of Aromatics
FPV	final plant value
FRV	final residue value
GI	gastrointestinal
GMATC	geometric maximum acceptable tissue concentration
HC	hazardous concentration
HCN	hydrogen cyanide
HLA	Harding Lawson Associates
HPLC	high performance liquid chromatography
IFEM	integrated fate and effects model
IRIS	Integrated Risk Information System
LL	less-than-lifetime
LOAEL	lowest observed adverse effect level
LOEC	lowest observed effect concentration
MATC	maximum acceptable toxicant concentration
MATissueC	maximum acceptable tissue concentration

MF	modifying factor
NAS	National Academy of Science
NOAA	National Oceanic and Atmospheric Administration
NOAEL	no observed adverse effect level
NOEC	no observed effect concentration
PAH	polycyclic aromatic hydrocarbon
PCB	polychlorinated biphenyl
PI	prediction interval
QSAR	quantitative structure activity relationship
RAF	relative absorption factor
RfD	reference dose
RMA	Rocky Mountain Arsenal
SETAC	Society of Environmental Toxicology and Chemistry
SF	safety factor
SQC	sediment quality criteria
SS	steady state
SSLC	Species Screening Level Concentration
TQ	toxicity quotient
TRV	terrestrial reference value
UF	uncertainty factor
WQC	water quality criteria

Index

Abiotic domain, 190
Accumulation, 28. *See also*
  Bioaccumulation
ACRs. *See* Acute-to-chronic ratios
Acute-to-chronic extrapolation,
  96, 99–100, 213, 216, 218
Acute-to-chronic ratios (ACRs),
  52–53, 54, 57
  derivation of, 82
  modeling compared to, 57–60
  uncertainty factors and, 82–83
Acute-to-chronic uncertainty
  factors, 50–60, 84–87, 89,
  100, 217–218
  acute-to-chronic ratios and,
  52–53, 54, 57
  modeling of, 53–57
Acute response, 11
Acute toxicity testing, 63, 189
Additive effects, 88, 231, 240
AET. *See* Apparent effects
  threshold
AFs. *See* Application factors
Aldrin, 136–137, 141
  in aquatic food web, 126–129,
  132
  hazard index for, 120
  in mammalian species, 133
  in terrestrial food web, 122–125
Aliphatics, 52. *See also* specific
  types
Ambient water quality criteria,
  143
o-Aminobenzoic acid, 29
p-Aminobenzoic acid, 29

Anilines, 238. *See also* specific
  types
Antibiotics, 231. *See also* specific
  types
Apparent effects threshold (AET)
  approach, 144–149, 158
Application factors (AFs), 11, 51,
  52–53, 188
Aromatics, 238. *See also* specific
  types
Arsenic, 120, 126–129, 132, 133

Barbiturate-induced sleeping
  time, 40
BCF. *See* Bioconcentration factors
Behavioral profiles, 8
Benz(a)anthracene, 157
Benzenes, 11, 29, 31, 32, 52. *See
  also* specific types
Benzo(a)anthracene, 153
Benzo(a)pyrene, 153, 157
Beryllium, 52
Bioaccumulation, 122–125. *See
  also* Accumulation
  chemical-specific MATCs and,
  165, 166
  models of, 23–24
  sediment quality criteria and,
  153, 162
Bioavailability, 114–115
Biochemical effects, 115–116
Bioconcentration factors (BCF).
  See also Concentrations
  chemical-specific MATCs and,
  172

247